INTRODUCTION TO FORMAL LANGUAGES

György E. Révész

IBM T.J. Watson Research Center

Dover Publications, Inc., New York

To my sons,
Szilárd, Tamás, and Peter

Published in Canada by General Publishing Company, Ltd., 30 Lesmill Road, Don Mills, Toronto, Ontario.
Published in the United Kingdom by Constable & Company, Ltd., 3 The Lanchesters, 162–164 Fulham Palace Road, London W6 9ER.

This Dover edition, first published in 1991, is an unabridged and slightly corrected republication of the work first published by the McGraw-Hill Book Company, New York, in 1983 in the *McGraw-Hill Computer Science Series*.

Manufactured in the United States of America

Dover Publications, Inc.
31 East 2nd Street
Mineola, New York 11501

Library of Congress Cataloging-in-Publication Data

Révész, György E.
   Introduction to formal languages / György E. Révész.
      p.   cm.
   "Unabridged and slightly corrected republication of the work first published by the McGraw-Hill Book Company in 1983 in the McGraw-Hill computer science series"—T.p. verso.
   Includes bibliographical references and index.
   ISBN 0-486-66697-2 (pbk.)
   1. Formal languages.   I. Title.
QA267.3.R485   1991
511.3—dc20
                                                              91-11606
                                                              CIP

CONTENTS

PREFACE

Formal languages have been playing a fundamental role in the progress of computer science. Formal notations in mathematics, logic, and other sciences had been developed long before the advent of electronic computers. The evolution of formal notations was slow enough to allow the selection of those patterns which appeared to be the most favorable for their users, but the computer revolution has resulted in an unprecedented proliferation of artificial languages. Given time, I presume, natural selection would take effect, but for now we have to face this modern story of the Tower of Babel.

The purpose of a scientific theory is to bring order to the chaos, and this is what formal language theory is trying to do—with partial success. Unfortunately, the theory itself sometimes appears to be in serious disorder, and nobody knows for sure which way it will turn next time. One important feature, however, is apparent so far: there is a close relationship between formal grammars and other abstract notions used in computer science, such as automata and algorithms. Indeed, since the results in one theory can often be translated into another, it seems to be an arbitrary decision as to which interpretation is primary.

In this book, formal grammars are given preferential treatment because they are probably the most commonly known of the various theories among computer scientists. This is due to the success of context-free grammars in describing the syntax of programming languages. For this reason, I have also changed the proofs of some of the classical theorems which are usually shown with the aid of automata (e.g., Theorems 3.9 and 3.12). In this way, the introduction of the latter notion is deferred until it really becomes necessary; at the same time a more uniform treatment of the subject is presented. The connection between grammars and automata is also emphasized by the similarity of the notation for derivation and reduction, respectively.

I did not try to account for all relevant results. Instead, I wanted to give a fairly coherent picture of mainstream formal language theory, which, of course, cannot be totally unbiased. In some cases, I have included less commonly known results to indicate particular directions in the development of the theory. This is especially true with Chapter 10, which is the first comprehensive presentation of its specific topic. I would like to take this opportunity to express my thanks to Johnson M. Hart for his contribution to this book as coauthor of Chapter 10, of which he wrote Sections 10.1, 10.4, 10.5, and 10.6. His enthusiasm and encouragement with respect to the writing of the entire book is greatly appreciated.

Chapters 1 through 9 can be used as the primary text for graduate or senior undergraduate courses in formal language theory. Automata theory and complexity theory are not studied here in depth. Nevertheless, the text provides sufficient background in those areas that they can be explored more specifically in other courses later on. The same is true for computability theory, that is, the theory of algorithms. In the areas of compiler design and programming language semantics, I feel that my book can have a favorable influence on the way of thinking about these subjects, though less directly.

Some of the theorems, e.g., Theorems 3.10, 3.12, 4.4, 5.2, and 8.10, may be skipped at the first reading of the book without impairing the understanding of the rest. Worked examples in the text usually form an integral part of the presentation. Exercises at the end of chapters or larger sections serve two purposes: they help in understanding the theory, and they illustrate some of its applications. I have made all efforts to simplify the proofs included in the book without compromising their mathematical rigor. A proof is not just a tool for convincing the skeptics but also an aid for better understanding the true nature of the result. I have therefore refrained from presenting theorems without proofs except in a very few cases.

György Révész

CHAPTER ONE

THE NOTION OF FORMAL LANGUAGE

1.1 BASIC CONCEPTS AND NOTATIONS

A finite nonvoid set of arbitrary symbols (such as the letters used for writing in some natural language or the characters available on the keyboard of a typewriter) is called a *finite alphabet* and will usually be denoted by V. The elements of V are called *letters* or *symbols* and the finite strings of letters are called *words* over V. The set of all words over V is denoted by V^*. The *empty word*, which contains no letters, will be denoted by λ and is considered to be in V^* for every V.

Two words written in one is called the *catenation* of the given words. The catenation of words is an associative operation, but in general it is noncommutative. Thus, if P and Q are in V^* then their catenation PQ is usually different from QP except when V contains only one letter. But for every P, Q, and R in V^* the catenation of PQ and R is the same as the catenation of P and QR. Therefore, the resulting word can be written as PQR without parentheses. The set V^* is obviously closed with respect to catenation, that is, the result PQ is always in V^* whenever both P and Q are in V^*. The empty word λ plays the role of the unit element for catenation, namely $\lambda P = P\lambda = P$ for every P. (We shall assume that P is in V^* for some given V even if we do not mention that.)

The *length* of P denoted by $|P|$ is simply the number of letters of P. Hence $|\lambda| = 0$ and $|PQ| = |P| + |Q|$ for every P and Q. Two words are equal if one is a letter by letter copy of the other. The word P is a *part* of Q if there are words P_1 and P_2 such that $Q = P_1 P P_2$. Further, if $P \neq \lambda$ and $P \neq Q$, then it is a proper part of Q and if $P_1 = \lambda$ or $P_2 = \lambda$ then P is a *head* (initial part) or a *tail* of Q, respectively.

For a positive integer i and for an arbitrary word P we denote by P^i the *i-times iterated catenation* of P (that is, i copies of P written in one word). By convention $P^0 = \lambda$ for every P. If, for example, $P = ab$, then $P^3 = ababab$. (Note that we need parentheses to distinguish $(ab)^3 = ababab$ from $ab^3 = abbb$.)

The *mirror image* of P, denoted by P^{-1}, is the word obtained by writing the letters of P in the reverse order. Thus, if $P = a_1 a_2 \cdots a_n$ then $P^{-1} = a_n a_{n-1} \cdots a_1$. Clearly, $(P^{-1})^{-1} = P$ and $(P^{-1})^i = (P^i)^{-1}$ for $i = 0, 1, \ldots$.

An arbitrary set of words of V^* is called a *language* and is usually denoted by L. The empty language containing no words at all is denoted by $\varnothing$. It should not be confused with the language $\{\lambda\}$ containing a single word λ. The set V^* without λ is denoted by V^+. A language $L \subseteq V^*$ is finite if it contains a finite number of words, otherwise it is infinite. The complete language V^* is always denumerably infinite. (The set of all subsets of V^*, that is, the set of languages over a finite alphabet, is nondenumerable.)

The above notion of language is fairly general but not extremely practical. It includes all written natural languages as well as the artificial ones, but it does not tell us how to define a particular language. Naturally, we want finite—and possibly concise—descriptions for the mostly infinite languages we are dealing with. In some cases we may have a finite characterization via some simple property. If, for instance, $V = \{a, b\}$ then

$$L_1 = \{a, b, \lambda\}$$

$$L_2 = \{a^i b^i | i = 0, 1, \ldots\}$$

$$L_3 = \{PP^{-1} | P \in V^*\}$$

$$L_4 = \{a^{n^2} | n = 1, 2, \ldots\}$$

are all well-defined languages. Or let $N_a(P)$ and $N_b(P)$ denote the number of occurrences of a and b, respectively, in P. Then

$$L_5 = \{P | P \in \{a, b\}^+ \quad \text{and} \quad N_a(P) = N_b(P)\}$$

is also well-defined. But we need other, more specific tools to define more realistic languages. For this purpose the notion of generative grammar will be introduced as follows:

Definition 1.1 A *generative grammar* G is an ordered fourtuple (V_N, V_T, S, F) where V_N and V_T are finite alphabets with $V_N \cap V_T = \varnothing$, S is a distinguished symbol of V_N, and F is a finite set of ordered pairs (P, Q) such that P and Q are in $(V_N \cup V_T)^*$ and P contains at least one symbol from V_N.

The symbols of V_N are called *nonterminal* symbols or *variables* and will usually be denoted by capital letters. The symbols of V_T are called *terminal* symbols and will be denoted by small letters. According to Definition 1.1 the sets V_N and V_T are disjoint in every grammar. The nonterminal symbol S is called the *initial symbol* and is used to start the derivations of the words of the language.

The ordered pairs in F are called *rewriting rules* or *productions* and will be written in the form $P \to Q$ where the symbol $\to$ is, of course, not in $V_N \cup V_T$. Productions are used to derive new words from given ones by replacing a part equal to the left-hand side of a rule by the right-hand side of the same rule. The precise definitions are given below.

Definition 1.2: Derivation in one step Given a grammar $G = (V_N, V_T, S, F)$ and two words $X, Y \in (V_N \cup V_T)^*$, we say that Y is derivable from X in one step, in symbols $X \underset{G}{\Rightarrow} Y$, iff there are words P_1 and P_2 in $(V_N \cup V_T)^*$ and a production $P \to Q$ in F such that $X = P_1 P P_2$ and $Y = P_1 Q P_2$.

Definition 1.3: Derivation Given a grammar $G = (V_N, V_T, S, F)$ and two words X, Y in $(V_N \cup V_T)^*$, we say that Y is derivable from X, in symbols $X \underset{G}{\overset{*}{\Rightarrow}} Y$, iff $X = Y$ or there is some word Z in $(V_N \cup V_T)^*$ such that $X \underset{G}{\overset{*}{\Rightarrow}} Z$ and $Z \underset{G}{\Rightarrow} Y$.

In other words, the relation $\underset{G}{\overset{*}{\Rightarrow}}$ is the reflexive and transitive closure of $\underset{G}{\Rightarrow}$. Occasionally, we shall use the transitive closure, denoted by $\underset{G}{\overset{+}{\Rightarrow}}$, which involves at least one step. The subscript G will usually be omitted when the context makes it clear which grammar is used.

Definition 1.4 The language generated by G is defined as

$$L(G) = \left\{ P \mid S \underset{G}{\overset{*}{\Rightarrow}} P \quad \text{and} \quad P \in V_T^* \right\}$$

or in other words

$$L(G) = \left\{ P \mid S \underset{G}{\overset{*}{\Rightarrow}} P \right\} \cap V_T^*$$

This means that the language generated by G contains exactly those words which are derivable from the initial symbol S and contain only terminal symbols.

Nonterminal symbols are used only as intermediary (auxiliary) symbols in the course of derivations. A derivation terminates when no more nonterminals are left in the word. (Note that according to Definition 1.1 the left-hand side of

4 INTRODUCTION TO FORMAL LANGUAGES

a rewriting rule must contain at least one nonterminal symbol.) A derivation aborts if there are nonterminals left in the word but there is no rewriting rule in F that can be applied to it.

Example 1.1 Let $G = (V_N, V_T, S, F)$ be a grammar where $V_N = \{S, A, B\}$, $V_T = \{a, b\}$, and the rules in F are as follows:

$$S \to aB \qquad S \to bA$$

$$A \to a \qquad A \to aS \qquad A \to bAA$$

$$B \to b \qquad B \to bS \qquad B \to aBB$$

We will show that this grammar generates the language (see L_5 above)

$$L = \{P \,|\, P \in \{a, b\}^+ \quad \text{and} \quad N_a(P) = N_b(P)\}$$

PROOF It is easy to see that in each word of $L(G)$ the number of a's must be the same as that of b's. Namely, in every word derivable from S the sum of the a's and A's is equal to that of the b's and B's. So we have established the inclusion $L(G) \subseteq L$.

The reverse inclusion will be shown by induction on the number of occurrences of a (or b) in the words of L.

Basis: Both ab and ba are in $L(G)$.

Induction step: Assume that every word in L having at most n occurrences of a does belong to $L(G)$, and let $P \in L$ have $n + 1$ occurrences of a ($|P| = 2n + 2$).

First consider the case when $P = a^i bX$ for some $i \geq 1$ and $X \in \{a, b\}^+$. If $i > 1$ then take the shortest head of X, denoted by U_1, such that $N_b(U_1) > N_a(U_1)$. Clearly, $U_1 = W_1 b$ for some word W_1 with $N_b(W_1) = N_a(W_1)$. (Note that W_1 may be λ.) This process can be repeated with the rest of X until we get the decomposition of P of the form

$$P = a^i bW_1 b \cdots W_{i-1} bW_i$$

where each W_j (for $j = 1, \dots, i$) is either λ or it is in L. For $W_j \neq \lambda$ the induction hypothesis gives us $S \overset{*}{\Rightarrow} W_j$. But for every $i \geq 1$ we have $S \overset{*}{\Rightarrow} a^i B^i$, $B \Rightarrow b$, and $B \Rightarrow bS$ in our grammar, which gives the result $S \overset{*}{\Rightarrow} P$.

Because of symmetry (the roles of a and b can simply be exchanged), the case $P = b^i aY$ need not be discussed separately; this completes the proof.

Example 1.2 Let $G = (\{S, X, Y\}, \{a, b, c\}, S, F)$ where the rules in F are:

$$S \to abc \qquad S \to aXbc$$

$$Xb \to bX \qquad Xc \to Ybcc$$

$$bY \to Yb \qquad aY \to aaX \qquad aY \to aa.$$

We show that this grammar generates the language

$$\{a^n b^n c^n \mid n = 1, 2, \ldots\}$$

PROOF First we show by induction on i that $S \overset{*}{\Rightarrow} a^i Xb^i c^i$ for every $i \geqslant 1$.
For $i = 1$ the assertion is trivial. Assume that $S \overset{*}{\Rightarrow} a^i Xb^i c^i$ is true for some $i \geqslant 1$. Then this derivation can be continued only by applying the rule $Xb \to bX$ i times, then using $Xc \to Ybcc$ and applying $bY \to Yb$ i times. This way we get $S \overset{*}{\Rightarrow} a^i Xb^i c^i \overset{*}{\Rightarrow} a^i Yb^{i+1} C^{i+1}$. Now we have two possibilities: the rule $aY \to aaX$ yields

$$S \overset{*}{\Rightarrow} a^{i+1} Xb^{i+1} c^{i+1}$$

which was to be shown first, while the rule $aY \to aa$ yields

$$S \overset{*}{\Rightarrow} a^{i+1} b^{i+1} c^{i+1}$$

Hence, the latter also holds true for every $i \geqslant 1$.

It can be seen, further, that no other words are derivable in this grammar, since we have always exactly one rule that can be applied except for the last step in the above derivation. But the application of the rule $aY \to aa$ will always terminate the derivation so it can be continued only in one way, which completes the proof.

1.2 THE CHOMSKY HIERARCHY OF LANGUAGES

As can be seen from the definitions given above, every grammar generates a unique language, but the same language can be generated by several grammars.

Definition 1.5 Two grammars are called *weakly equivalent* (or, more simply, *equivalent*) if they generate the same language.

Naturally, the question arises whether we can recognize equivalent grammars just by looking at them. In other words, is there any similarity between two equivalent grammars? Unfortunately, it is impossible to solve this problem in general. We can, however, classify our generative grammars on the basis of the forms of their production rules. The classification given below has been introduced by N. Chomsky, who is the founder of the whole theory.

Definition 1.6 A generative grammar $G = (V_N, V_T, S, F)$ is said to be of *type i* if it satisfies the corresponding restrictions in this list:

$i = 0$: No restrictions.

$i = 1$: Every rewriting rule in F has form $Q_1 A Q_2 \rightarrow Q_1 P Q_2$, with Q_1, Q_2, and P in $(V_N \cup V_T)^*$, $A \in V_N$, and $P \neq \lambda$, except possibly for the rule $S \rightarrow \lambda$, which may occur in F, in which case S does not occur on the right-hand sides of the rules.

$i = 2$: Every rule in F has form $A \rightarrow P$, where $A \in V_N$ and $P \in (V_N \cup V_T)^*$.

$i = 3$: Every rule in F has form either $A \rightarrow PB$ or $A \rightarrow P$, where $A, B \in V_N$ and $P \in V_T^*$.

A language is said to be of *type i* if it is generated by a *type i* grammar. The class (or family) of type i languages is denoted by $\mathcal{L}_i$.

Type 0 grammars are often called *phrase structure* grammars, which refers to their linguistical origin. Type 1 grammars are called *context-sensitive*, since each of their rules allows for replacing an occurrence of a nonterminal symbol A by the corresponding word P only in the context Q_1, Q_2. On the contrary, the rules of a type 2 grammar are called *context-free*, as they allow for such replacements in any context. Type 3 grammars are called *regular* or *finite state* (the latter expression refers to their connection with finite automata, which will be discussed later).

We can illustrate the linguistic background with this example. Consider the following sentence:

The tall boy quickly answered the mathematical question.

Now, let us denote a sentence by S, a noun phrase by A, a verb phrase by B, an article by C, an adjective by D, a noun by E, an adverb by G, and a verb by F. Then take the following productions:

$$S \rightarrow A + B \qquad A \rightarrow C + D + E$$
$$B \rightarrow G + B \qquad B \rightarrow F + A$$
$$C \rightarrow \text{the} \qquad D \rightarrow \text{mathematical}$$
$$D \rightarrow \text{tall} \qquad E \rightarrow \text{boy}$$
$$E \rightarrow \text{question} \qquad G \rightarrow \text{quickly}$$
$$F \rightarrow \text{answered}$$

where the $+$ sign represents the space as a terminal symbol. It is easy to see that the given sentence can be derived from S with the aid of these rules. (Note that the words "tall" and "mathematical" can be exchanged in the sentence without destroying its syntax, though its meaning will be "slightly" different.) Thus, we can derive syntactically correct sentences with the aid of such grammars. But the English language, just like any other natural language, contains much more complicated sentences, too. It is, therefore, impossible to describe the complete syntax of a natural language using only such simple rules. Nevertheless, generative grammars are useful tools for studying the basic structures of sentences of natural languages. Artificial languages, on the other hand, can be defined in such a way that they are generated by generative grammars.

Returning to the classification of formal languages, it is obvious from Definition 1.6 that every type 3 language is also type 2 and every type 1 language is also type 0. It is also trivial that they are all type 0 at the same time.

This means that

$$\mathcal{L}_3 \subseteq \mathcal{L}_2 \subseteq \mathcal{L}_0 \quad \text{and} \quad \mathcal{L}_1 \subseteq \mathcal{L}_0$$

but the relationship between $\mathcal{L}_2$ and $\mathcal{L}_1$ is not obvious because of the restrictions related to the empty word λ. Later we shall see that the inclusion $\mathcal{L}_2 \subseteq \mathcal{L}_1$ is also true and that we have here a proper hierarchy of language classes

$$\mathcal{L}_3 \subset \mathcal{L}_2 \subset \mathcal{L}_1 \subset \mathcal{L}_0$$

where all the inclusions are proper. The latter assertion is hardly trivial, since every language can be generated by several grammars which need not necessarily be of the same type.

As far as applications are concerned, context-free grammars are widely used in computer science for defining programming languages like FORTRAN, ALGOL, COBOL, PASCAL, etc. Context-sensitive grammars are also used for similar purposes when context-dependent features are to be modeled. Type 3 languages are closely related to the theory of finite automata, which will not be discussed extensively in this book. We shall cover only those aspects of automata theory that are relevant to the theory of formal languages. (The interested reader may find a few titles on automata theory in our references as well.) Unrestricted phrase structure, i.e., type 0, grammars are less useful from the practical point of view, since they are, as we shall see later, very difficult to handle.

EXERCISES

1.1 Show that the grammar

$$G = (\{S\}, \{a, b\}, S, \{S \rightarrow \lambda, S \rightarrow aSb\})$$

generates the language

$$L = \{a^i b^i \mid i = 0, 1, \ldots\}$$

1.2 Find a context-free grammar to generate each of the following languages:
   a) $L = \langle a^i b^j | i = 0, 1, \ldots, j = 0, 1, \ldots, \text{ and } j \geqslant i \rangle$
   b) $L = \langle a^i b^j a^j b^i | i = 0, 1, \ldots, j = 0, 1, \ldots \rangle$
   c) $L = \langle a^i b^i a^j b^j | i = 0, 1, \ldots, j = 0, 1, \ldots \rangle$
   d) $L = \langle a^i b^i | i = 0, 1, \ldots \rangle \cup \langle b^j a^j | j = 0, 1, \ldots \rangle$
   e) $L = \langle PP^{-1} | P \in \langle a, b \rangle^* \rangle$
   f) $L = \langle a^i b^j c^{i+j} | i = 0, 1, \ldots, j = 0, 1, \ldots \rangle$

1.3 Show that the language given in Example 1.1 can be generated also by the grammar

$$G = (\langle S \rangle, \langle a, b \rangle, S, F)$$

where the rules in F are these

$$S \to ab \qquad S \to ba \qquad S \to SS$$

$$S \to aSb \qquad S \to bSa$$

1.4 Show that the language generated by the grammar

$$G = (\langle A, B, S \rangle, \langle a, b \rangle, S, F)$$

with F consisting of the rules

$$S \to aAB \qquad bB \to a \qquad Ab \to SBb$$

$$Aa \to SaB \qquad B \to SA \qquad B \to ab$$

is empty. *Hint:* Observe that at least one nonterminal symbol always remains in every word derivable from S.

1.5 Convince yourself that the language

$$L = \left\{ a^{n^2} | n = 1, 2, \ldots \right\}$$

can be generated by the grammar

$$G = (\langle S, A, B, C, D, E \rangle, \langle a \rangle, S, F)$$

where the rules in F are

$$S \to a \qquad S \to CD \qquad C \to ACB$$

$$C \to AB \qquad AB \to aBA \qquad Aa \to aA$$

$$Ba \to aB \qquad AD \to Da \qquad BD \to Ea$$

$$BE \to Ea \qquad E \to a$$

Hint: First try to derive $A^n B^n D$, then $a^{n^2} B^n A^n D$, which yields $a^{(n+1)^2}$. Of course, it is necessary to show that no other terminal strings are derivable.

1.6 Does the grammar

$$G = (\langle S, A, B \rangle, \langle a, b, c, d \rangle, S, \langle S \to AB, A \to Ab, A \to a, B \to cB, B \to d \rangle)$$

generate a type 3 language? Be careful; a language can be generated by several grammars.

CHAPTER TWO

OPERATIONS ON LANGUAGES

2.1 DEFINITIONS OF OPERATIONS ON LANGUAGES

Languages have been defined above as sets of words, and thus the usual set-theoretical operations can be defined on them directly. Hence

$$L_1 \cup L_2 = \{P \mid P \in L_1 \text{ or } P \in L_2\}$$
$$L_1 \cap L_2 = \{P \mid P_1 \in L_1 \text{ and } P \in L_2\}$$
$$L_1 - L_2 = \{P \mid P \in L_1 \text{ and } P \notin L_2\}$$

The complement of L with respect to V^* for $L \subseteq V^*$ is defined as

$$\bar{L} = V^* - L$$

Also, the catenation of words can be extended to languages easily as

$$L_1 L_2 = \{P_1 P_2 \mid P_1 \in L_1 \text{ and } P_2 \in L_2\}$$

This is then used to define the iteration L^i for every positive integer i. By convention let $L^0 = \{\lambda\}$. We have here the identities

$$\emptyset L = L\emptyset = \emptyset$$

and

$$\{\lambda\}L = L\{\lambda\} = L$$

for any language L. The closure of the iteration denoted by L^* is defined as

$$L^* = \bigcup_{i \geq 0} L^i$$

This notation is in accordance with the notation V^* if we consider the alphabet V as a finite language containing one letter words. Defining L^+ as

$$L^+ = \bigcup_{i \geqslant 1} L^i$$

we obviously have the relationship

$$L^+ = L^* \qquad \text{if } \lambda \in L$$
$$L^+ = L^* - \{\lambda\} \qquad \text{if } \lambda \notin L$$

The mirror image of a language is defined simply as

$$L^{-1} = \{P | P^{-1} \in L\}$$

for which

$$(L^{-1})^{-1} = L$$

and

$$(L^{-1})^i = (L^i)^{-1} \qquad i = 0, 1, \ldots$$

also hold.

The head of a language $L \subseteq V^*$ is defined as

$$\text{HEAD}(L) = \{P | P \in V^* \text{ and there is some } Q \in V^* \text{ such that } PQ \in L\}$$

This means that the head of a language consists of those words over V which can be completed in such a way that the complete word belongs to L. Thus, by definition

$$L \subseteq \text{HEAD}(L)$$

for all L.

Another important operation is homomorphism that is commutable with catenation in the following sense. For every pair of words $P, Q \in V^*$ the homomorphic image of their catenation, that is, $h(PQ)$, is the same as the catenation of their images $h(P)$ and $h(Q)$. In short

$$h(PQ) = h(P)h(Q)$$

for all P and Q.

Definition 2.1 Let V_1 and V_2 be two alphabets. Then a mapping h of V_1^* into V_2^* is called a *homomorphism* if it has these two properties:

1) h is unique, that is, for every P in V_1^* there is exactly one word W in V_2^* with $W = h(P)$.
2) $h(PQ) = h(P)h(Q)$ for all P, Q in V_1^*.

These two properties imply that $h(\lambda) = \lambda$. Namely, for every P in V_1^*

$$h(P) = h(\lambda P) = h(\lambda)h(P)$$

It is clear, further, that every homomorphism is completely defined whenever it is defined for the letters of V_1. Indeed, for every $P \in V_1^*$ with

$$P = a_1 a_2 \cdots a_n \quad (a_i \in V_1)$$

we have

$$h(P) = h(a_1)h(a_2) \cdots h(a_n)$$

Therefore, it is enough to specify the mapping h for the individual letters of V_1 and this will be extended automatically to V^*. A homomorphism is called λ-free if for all P, $P \neq \lambda$ implies $h(P) \neq \lambda$.

The homomorphic image of a language will then be defined as

$$h(L) = \{W \in V_2^* \mid W = h(P) \text{ for some } P \in L\}$$

A homomorphism h is called an *isomorphism* if for all P and Q in V_1^* $h(P) = h(Q)$ implies $P = Q$. A special case of isomorphism arises when the images of the letters of V_1 are letters in V_2. Such an isomorphism is merely a transliteration from V_1 into V_2. Two languages which differ only in the graphical (i.e., physical) representation of their letters are considered the same.

A good example of the isomorphism is the so-called binary coded decimal representation of the integers where

$$V_1 = \{0, 1, 2, \ldots, 9\}, \qquad V_2 = \{0, 1\}$$

and $\qquad h(0) = 0000, \qquad h(1) = 0001, \ldots, h(9) = 1001$

Homomorphism is very useful in our theory, as can be seen in the proof of the following normal form theorem.

Theorem 2.1 For every grammar $G = (V_N, V_T, S, F)$ we can give an equivalent grammar $G' = (V_N', V_T, S, F')$ of the same type such that terminal letters do not occur on the left-hand sides of the rules in F'.

PROOF For type 3 and type 2 grammars we have nothing to prove. If G is of type 1 or type 0 then we introduce to each terminal letter $a_i \in V_T$ a new nonterminal $A_i \notin V_N$. Then let $V_N' = V_N \cup \{A_1, \ldots, A_k\}$ and let F' be obtained from F by replacing a_i by A_i on both sides of the rules and adding the rules $A_i \to a_i$ for $i = 1, 2, \ldots, k$ where k is the number of the terminal letters. Now, we can see that

$$L(G) \subseteq L(G')$$

For $P = a_{i_1} a_{i_2} \cdots a_{i_n} \in L(G)$ we can derive $A_{i_1} A_{i_2} \cdots A_{i_n}$ in G'. But using the rules $A_i \to a_i$ we get $P \in L(G')$. For $P = \lambda$ it is also clear that $\lambda \in L(G)$ implies $\lambda \in L(G')$. In order to show the reverse inclusion

$$L(G') \subseteq L(G)$$

we define the homomorphism h from $(V'_N \cup V_T)^*$ into $(V_N \cup V_T)^*$ such that

1) $h(A_i) = a_i$ $i = 1, 2, \ldots, k$
2) $h(x) = x$ for $x \in V_N \cup V_T$

Hence for each pair of words P, Q in $(V'_N \cup V_T)^*$ the relation $P \underset{G'}{\Rightarrow} Q$ implies $h(P) \underset{G}{\overset{*}{\Rightarrow}} h(Q)$. For if Q is derived from P by using one of the rules $A_i \to a_i$ then $h(P) = h(Q)$. Otherwise $P \underset{G'}{\Rightarrow} Q$ involves the application of a rule obtained from a rule in F and thus, $h(P) \underset{G}{\Rightarrow} h(Q)$. Therefore, $P \underset{G'}{\overset{*}{\Rightarrow}} Q$ implies $h(P) \underset{G}{\overset{*}{\Rightarrow}} h(Q)$, and in particular, $S \underset{G'}{\overset{*}{\Rightarrow}} P$ for $P \in V_T^*$ implies $S = h(S) \underset{G}{\overset{*}{\Rightarrow}} h(P) = P$ which means that $L(G') \subseteq L(G)$ and this completes the proof.

REMARK For type 0 and type 1 grammars the above proof yields a normal form where terminal letters occur only in rewriting rules with the form $A \to a$, $A \in V_N$, $a \in V_T$. This stronger normal form can be obtained also for type 2 grammars. For type 3 grammars, however, the construction does not work since we cannot have more than one nonterminal on the right-hand side of a rule in this case.

2.2 CLOSURE PROPERTIES OF LANGUAGE CLASSES

The set union, catenation, and the closure of the iteration are called *regular operations*. These operations have some very useful properties. First of all, we shall prove that for each class $\mathfrak{L}_i$ ($i = 0, 1, 2, 3$) these operations do not lead outside of the given class.

Theorem 2.2 Each of the language classes $\mathfrak{L}_i$ ($i = 0, 1, 2, 3$) is closed under the regular operations.

PROOF Let L and L' be two languages of type i. This means that there are two type i grammars $G = (V_N, V_T, S, F)$ and $G' = (V'_N, V'_T, S', F')$ such that $L = L(G)$ and $L' = L(G')$. According to Theorem 2.1 we can assume that terminal symbols do not occur on the left-hand sides of the rules in F and F'. Moreover, we can always achieve that V_N and V'_N be disjoint, $V_N \cap V'_N = \varnothing$, since the choice of the particular symbols representing the nonterminals does not influence the generated language.

The proof will be given separately for each operation.

UNION Let first $i = 0, 2, 3$ and $S_0 \notin V_N \cup V'_N$ be a newly introduced nonterminal symbol. Then the grammar

$$G_u = \left(V_N \cup V'_N \cup \{S_0\}, V_T \cup V'_T, S_0, F \cup F' \cup \{S_0 \to S, S_0 \to S'\} \right)$$

generates the language $L \cup L'$ and is of type i provided that both G and G' are of type i. The second part of the assertion is trivial. For the first part it is obvious that

$$L \cup L' \subseteq L(G_u)$$

To prove the reverse inclusion

$$L(G_u) \subseteq L \cup L'$$

we only have to note that $V_N \cap V'_N = \varnothing$ and thus, the application of the rule $S_0 \to S$ prevents the application of the rules of F' in the rest of the derivation. Similarly, if a derivation starts with $S_0 \to S'$ then the rules of F cannot be applied in that derivation.

Now, let $i = 1$ and $\lambda \notin L \cup L'$. Then G_u constructed as above will do.

For $i = 1$ and $\lambda \in L \cup L'$ take first $L_1 = L - \{\lambda\}$ and $L_2 = L' - \{\lambda\}$ and construct G_u as above with L_1 and L_2 in place of L and L', respectively. Then introduce one more nonterminal symbol S_1 and include the rules $S_1 \to S_0$ and $S_1 \to \lambda$.

CATENATION Let first $i = 0, 2$ and $S_0 \notin V_N \cup V'_N$. Then the language LL' can be generated by the grammar

$$G_c = \left(V_N \cup V'_N \cup \{S_0\}, V_T \cup V'_T, S_0, F \cup F' \cup \{S_0 \to SS'\} \right)$$

which is of the same type as G and G' are. Again it is obvious that $LL' \subseteq L(G_c)$. To show the reverse inclusion consider a derivation $S_0 \underset{G_c}{\overset{*}{\Rightarrow}} P$ of some P in $(V_T \cup V'_T)^*$. For each step in this derivation

$$S_0 \Rightarrow P_1 \Rightarrow P_2 \Rightarrow \cdots P_m = P$$

we can show by induction on j that $P_j = Q_j Q'_j$ for some Q_j and Q'_j such that

$$S \underset{G}{\overset{*}{\Rightarrow}} Q_j \quad \text{and} \quad S' \underset{G}{\overset{*}{\Rightarrow}} Q'_j$$

For $j = 1$ this is trivial since $P_1 = SS'$ must be the case. But, if the assertion is true for some P_j then it must be true also for P_{j+1}. This follows from the fact that $V_N \cap V'_N = \varnothing$ and that terminal symbols do not occur on the left-hand sides of the rules. (Either a part of Q_j or else a part of Q'_j is rewritten to obtain P_{j+1}.) Hence, each word in $L(G_c)$ is also in LL'.

Now, let $i = 1$ and $\lambda \notin LL'$. Then the grammar G_c constructed as above will do.

For $i = 1$ and $\lambda \in LL'$ take first $L_1 = L - \{\lambda\}$ and $L_2 = L' - \{\lambda\}$ and construct G_c as before with L_1 and L_2 in place of L and L', respectively. The language LL' will be identical to one of the three languages

$$L_1 L_2 \cup L_2 \qquad L_1 L_2 \cup L_1 \qquad L_1 L_2 \cup L_1 \cup L_2 \cup \{\lambda\}$$

depending on whether λ is only in L and not in L', or conversely, or else in both of them. In any case $LL' \in \mathcal{L}_1$ follows from $L_1 L_2 \in \mathcal{L}_1$ and from the fact that $\mathcal{L}_1$ is closed under the union.

For $i = 3$ let F_1 be obtained from F by replacing each rule of the form $A \to P$, $A \in V_N$, $P \in V_T^*$ by $A \to PS'$, while the other rules are left unchanged. Then the grammar

$$G_c = \left(V_N \cup V_N', V_T \cup V_T', S, F_1 \cup F'\right)$$

generates the language LL' and is of type 3. Surely, a derivation in G_c can only use the rules F_1 until a rule of the form $A \to PS'$ is applied. From that point on it can only use the rules F'. Hence, it produces first some word QS' with $Q \in L$ and then QQ' with $Q' \in L'$. (Note that in each step of a derivation in a type 3 grammar we have a unique nonterminal symbol that can be replaced.) So we have established that $L(G_c) \subseteq LL'$. It is easy to see that also $LL' \subseteq L(G_c)$ holds.

ITERATION CLOSURE First let $i = 2$ and $S_0 \notin V_N$. Then the grammar

$$G_* = \left(V_N \cup \{S_0\}, V_T, S_0, F \cup \{S_0 \to \lambda, S_0 \to SS_0\}\right)$$

obviously generates L^*.

For $i = 3$ let F_* be defined such that $A \to PS$ is in F_* iff $A \to P$, with $A \in V_N$ and $P \in V_T^*$, is in F. Then the grammar

$$G_* = \left(V_N \cup \{S_0\}, V_T, S_0, F_* \cup F \cup \{S_0 \to \lambda, S_0 \to S\}\right)$$

generates the language L^*.

Let next $i = 0, 1$ and $\lambda \notin L$. Then the grammar

$$G_* = \left(V_N \cup \{S_0, S_1\}, V_T, S_0, F \cup \{S_0 \to \lambda, S_0 \to S, S_0 \to S_1 S\}\right.$$
$$\left. \cup \{S_1 a \to S_1 Sa | a \in V_T\} \cup \{S_1 a \to Sa | a \in V_T\}\right)$$

where $S_0, S_1 \notin V_N$ generates the language L^*.

It is easy to see that $L^* \subseteq L(G_*)$. The difficult part is to show that no other terminal word is in $L(G_*)$. Consider a derivation

$$S_0 \underset{G_*}{\Rightarrow} P_1 \underset{G_*}{\Rightarrow} P_2 \cdots \underset{G_*}{\Rightarrow} P_n = P$$

with some $P \in V_T^*$. If in the first step the rule $S_0 \to \lambda$ is applied then $n = 1$ and $P_1 = P_n = \lambda \in L^*$. If $P_1 = S$ then $P \in L$ must be the case. Otherwise we have $P_1 = S_1 S$ and for every j $(1 \leqslant j \leqslant n)$ we can show by induction on j that each P_j is either of the two forms:

1) $S_1 Q_1 \cdots Q_k$ with $k \geqslant 1$ such that

$$S \underset{G}{\overset{*}{\Rightarrow}} Q_m \qquad \text{for } m = 1, \ldots, k$$

and each of the words $Q_2, \ldots, Q_k$ begins with a terminal symbol.

2) $Q_0 Q_1 \cdots Q_k$ with $k \geqslant 1$ such that

$$S \underset{G}{\overset{*}{\Rightarrow}} Q_m \qquad \text{for } m = 0, 1, \ldots, k$$

and each of the words $Q_1, \ldots, Q_k$ begins with a terminal symbol.

The form of $P_1 = S_1 S$ clearly corresponds to form 1. By examining each rule of G_* (taking into account that terminal symbols do not occur on the left-hand sides of the rules in F) we can show that P_{j+1} is of form 1 or 2 whenever this is true for P_j. This implies that $L(G_*) \subseteq L^*$.

Finally, if $i = 0, 1$ and $\lambda \in L$ then we construct a grammar G_1 with $L(G_1) = L - \{\lambda\}$. For $i = 1$ this can be done simply by leaving out the rule $S \to \lambda$. For $i = 0$ we replace each rule of the form $P \to \lambda$ by the rules $PX \to X$ and $XP \to X$ for all X in $V_N \cup V_T$. The type of G_1 is the same as that of G and $(L - \{\lambda\})^* = L^*$. Therefore, it is enough to give the construction for the case $\lambda \notin L$ and this completes the proof.

The families of languages $\mathcal{L}_i$ have many other properties in common besides their closure properties under the regular operations. A systematic study of these common properties has led to the development of the theory of abstract families of languages (AFL) which is the main subject of the book "Algebraic and Automata-Theoretic Properties of Formal Languages", by S. Ginsburg. An abstract family of languages is defined by axioms describing its closure properties under the operations in question. This makes it possible to prove general theorems about all families of languages satisfying these axioms. There are, of course, other classes of languages which satisfy the AFL axioms but the classes $\mathcal{L}_i$ in the Chomsky hierarchy seem to be the most important ones. In the present book we do not follow this axiomatic theory, so that we may rely upon our intuition about generative grammars.

Closure properties have not only theoretical but also practical aspects. If, for instance, we can define some of the subsets of a language L such that each subset is generated by a context-free grammar and their union is L, then we can construct a context-free grammar to generate L. Such an application is illustrated in the following example.

Example 2.1 Let V_1 and V_2 denote the English alphabet and the decimal digits, respectively. Suppose we want to develop a type 3 grammar that generates the language whose words each contain either letters only or digits only or else starts with letters and continues with digits. If we include the empty word as well then this language can be defined as

$$L = V_1^* \cup V_2^* \cup V_1^* V_2^* = V_1^* V_2^*$$

Let $S_1, S_2 \notin V_1 \cup V_2$ be introduced as nonterminal symbols and define two grammars $G_1 = (\{S_1\}, V_1, S_1, F_1)$ and $G_2 = (\{S_2\}, V_2, S_2, F_2)$ where the sets of rules are

$$F_1 = \{S_1 \rightarrow \lambda, S_1 \rightarrow aS_1, \ldots, S_1 \rightarrow zS_1\}$$

and

$$F_2 = \{S_2 \rightarrow \lambda, S_2 \rightarrow 0S_2, \ldots, S_2 \rightarrow 9S_2\}$$

These grammars are clearly of type 3 and $L(G_1) = V_1^*$, $L(G_2) = V_2^*$. Now, we define

$$F_1' = \{S_1 \rightarrow S_2, S_1 \rightarrow aS_1, \ldots, S_1 \rightarrow zS_1\}$$

and the grammar

$$G' = (\{S_1, S_2\}, V_1 \cup V_2, S_1, F_1' \cup F_2)$$

which is of type 3 and generates the language $V_1^* V_2^*$.

EXERCISES

2.1 Show that every finite language is in $\mathcal{L}_3$.

2.2 Prove that each of the families $\mathcal{L}_0$, $\mathcal{L}_2$, and $\mathcal{L}_3$ is closed under homomorphism and $\mathcal{L}_1$ is closed under λ-free homomorphism. *Hint:* Take homomorphic images on both sides of the rules.

2.3 Prove that if L is in $\mathcal{L}_3$, then HEAD(L) is also in $\mathcal{L}_3$. *Hint:* Cut off the ends of the rules in every possible way.

2.4 Show that each of the families $\mathcal{L}_0$, $\mathcal{L}_1$ and $\mathcal{L}_2$ are closed under mirror image. *Hint:* Take mirror images on both sides of the rules.

THREE

CONTEXT-FREE LANGUAGES

3.1 THE CHOMSKY NORMAL FORM

In the preceding chapter we have established some of the common properties of language families in the Chomsky hierarchy. The present chapter is devoted to context-free languages, which are probably the most interesting ones in view of the applications.

A great deal of the study of formal languages is concerned with various kinds of transformations on generative grammars. Especially, we are interested in transformations which render the form of the grammar simpler in some sense without changing the generated language. Here we shall prove first that each type 2 grammar can be brought to a form that satisfies the restrictions given in the definition of type 1 grammars. This way we can show that $\mathfrak{L}_2 \subseteq \mathfrak{L}_1$ holds, but the proper inclusion $\mathfrak{L}_2 \subset \mathfrak{L}_1$ remains to be shown later.

Theorem 3.1 To every context-free grammar G we can find an equivalent context-free grammar G' such that the right-hand sides of its rules are all different from λ except when $\lambda \in L(G)$ in which case $S' \to \lambda$ is the only rule with the right-hand side λ but then S' does not occur on the right-hand sides of the rules.

PROOF Given a context-free grammar $G = (V_N, V_T, S, F)$ we define the subsets of nonterminals $U_i \subseteq V_N$ as follows:

$$U_1 = \{ X | X \to \lambda \in F \}$$

$$U_{i+1} = U_i \cup \{ X | X \to P \in F \text{ for some } P \in U_i^* \} \qquad \text{for } i \geqslant 1$$

These sets form a nondecreasing sequence, that is, $U_i \subseteq U_{i+1}$ for all i, hence there is some index k for which $U_{k+1} = U_k$. (The set V_N is finite.) Obviously, $U_{k+j} = U_k$ with $j = 1, 2, \ldots$ also holds for this k. Denoting U_k simply by U it is easy to see that $X \overset{*}{\underset{G}{\Rightarrow}} \lambda$ holds for $X \in V_N$ iff X is in U. This means that $\lambda \in L(G)$ iff $S \in U$.

Let us define the set of rules F' this way: $X \to P'$ is in F' iff $P' \neq \lambda$ and there is a word $P \in (V_N \cup V_T)^*$ such that $X \to P$ is in F and P' can be obtained from P by erasing zero or more occurrences of one or more symbols belonging to U. (If, for instance, A and B are in U but $C \notin U$, then the rule $S \to ACAB$ gives rise to seven new rules: $S \to CAB$, $S \to ACB$, $S \to ACA$, $S \to CB$, $S \to CA$, $S \to AC$ and $S \to C$.) Then the language generated by $G' = (V_N, V_T, S, F')$ will be included in $L(G) - \{\lambda\}$ since each rule $X \to P' \in F'$ has the same effect as the corresponding rule $X \to P \in F$ combined with derivations of form $Z \overset{*}{\underset{G}{\Rightarrow}} \lambda$ with $Z \in U$. Conversely, if $S \overset{*}{\underset{G}{\Rightarrow}} P$ and $P \neq \lambda$ then there is also a derivation $S \overset{*}{\underset{G'}{\Rightarrow}} P$ since the need for the application of an $X \to \lambda$-like rule can be avoided by using the appropriate rule from F' instead.

So we have established that

$$L(G') = L(G) - \{\lambda\}$$

which is enough when $\lambda \notin L(G)$. If however, $\lambda \in L(G)$ then take the grammar

$$G_1 = (V_N \cup \{S_1\}, V_T, S_1, F' \cup \{S_1 \to \lambda, S_1 \to S\})$$

where $S_1 \notin V_N \cup V_T$, and this completes the proof.

The proof of this theorem includes a decision procedure for the question whether $\lambda \in L(G)$. The same question is trivial for type 1 languages but in general it turns out to be undecidable for type 0 languages. (See later in Chapter 7.)

As a corollary of Theorem 3.1 we can establish the relation

$$\mathcal{L}_2 \subseteq \mathcal{L}_1$$

which was left open in Chapter 1.

Definition 3.1 A grammar G is called λ-free if none of its rules has the right-hand side λ.

The proof of Theorem 3.1 shows that λ-rules (that is rules with form $X \to \lambda$) are not essential for context-free languages, only the word λ cannot be derived if we eliminate all λ-rules. In most cases we may, therefore, restrict ourselves to λ-free grammars. The proof of the theorem also works correctly for type 3 grammars so we can formulate the following corollary.

Corollary To every type 2 (type 3) grammar G we can construct a type 2 (type 3) grammar G' such that G' is λ-free and
$$L(G') = L(G) - \{\lambda\}$$

Definition 3.2 A context-free grammar is said to be in Chomsky normal form if each of its rules has either of the two forms

1) $X \rightarrow a, \qquad X \in V_N, a \in V_T$
2) $X \rightarrow YZ, \qquad X, Y, Z \in V_N$

Theorem 3.2 To every λ-free type 2 grammar $G = (V_N, V_T, S, F)$ one can find an equivalent grammar $G' = (V_N', V_T, S, F')$ in Chomsky normal form.

PROOF In virtue of Theorem 2.1 we can assume without the loss of generality that terminal symbols appear in F only in rules of the form $X \rightarrow a$, $X \in V_N$, $a \in V_T$.

All other rules have form $X \rightarrow P$ with $X \in V_N$ and $P \in V_N^*$. Replace each rule of the form
$$X \rightarrow Y_1 Y_2 \cdots Y_k \qquad (k \geqslant 3)$$
by a set of rules
$$X \rightarrow Y_1 Z_1$$
$$Z_1 \rightarrow Y_2 Z_2$$
$$\vdots$$
$$Z_{k-2} \rightarrow Y_{k-1} Y_k$$

where $Z_1, \ldots, Z_{k-2}$ are newly introduced nonterminal symbols. This way we get a grammar $G_1 = (V_N', V_T, S, F_1)$ where F_1 contains three kinds of rules

1) $X \rightarrow a \qquad X \in V_N', \qquad a \in V_T$
2) $X \rightarrow Y \qquad X, Y \in V_N'$
3) $X \rightarrow YZ \qquad X, Y, Z \in V_N'$

Note that V_N' contains the original symbols from V_N and the new nonterminals introduced separately for each rule that has been eliminated.

The $X \rightarrow Y$ like rules are called chain rules, and they are to be eliminated next. To this end we define for every $X \in V_N'$ the sets $U_i(X)$ as follows:
$$U_1(X) = \{X\}$$

and

$$U_{i+1}(X) = U_i(X) \cup \{Y \mid Y \to Z \in F_1 \text{ for some } Z \in U_i(X)\}$$

for $i \geqslant 1$. As V'_N is finite, there is an integer k such that $U_{k+j}(X) = U_k(X)$ for $j = 1, 2, \ldots$. Denoting $U_k(X)$ by $U(X)$ the relation $Y \overset{*}{\Rightarrow} X$ holds for $X, Y, \in V_N$ iff $Y \in U(X)$. (Thus, for any terminal symbol a the relation $a \in L(G)$ holds iff there is some $X \in V'_N$ such that $S \in U(X)$ and $X \to a$ is in F.) Now let us define F' in the following way:

1) $X \to a \in F'$ iff there is some $A \in V_N$ such that $X \in U(A)$ and $A \to a$ $\in F_1$
2) $X \to YZ \in F'$ iff there is some $A \in V_N$ such that $X \in U(A)$ and $A \to YZ \in F_1$

As can be seen from this construction $X \to a \in F'$ whenever $X \overset{*}{\underset{G_1}{\Rightarrow}} a$ holds and $X \underset{G'}{\Rightarrow} YZ$ whenever $X \overset{*}{\underset{G_1}{\Rightarrow}} A \underset{G_1}{\Rightarrow} YZ$ holds for some A. Therefore, to every derivation in G_1 there is a derivation in G' with the same result and vice versa which completes the proof.

REMARK When referring to the form of rules, the letters X, Y, Z, and so forth, denote arbitrary nonterminal symbols so they may occasionally represent the same symbol of the given grammar. (For example $A \to AB$ has the general form $X \to YZ$.) Of course, the letters appearing in a given example are each meant to be different.

Theorem 3.3 For every context-free grammar G it is decidable whether or not an arbitrary word P belongs to the language $L(G)$.

PROOF According to Theorem 3.1 it is enough to consider the case $P \neq \lambda$. Further, we can assume that G is in Chomsky normal form. Now, if P can be derived from S in k steps then $|P| \leqslant k + 1$. More exactly $A \to a$ type rules must be used $|P|$ times, hence $X \to YZ$ type rules are used $k - |P|$ times. This means that the length of P is exactly $1 + k - |P|$; hence,

$$k = 2|P| - 1$$

But we can produce a sequence of all words derivable from S in 1 step, 2 steps, $\ldots$, k steps. The number of such words is obviously finite for every finite k. Thus, we can choose $k = 2|P| - 1$ and see if P appears somewhere in this set; if not then we can be sure that $P \notin L(G)$.

Corollary For every context-free grammar G and every finite language L, both problems, whether $L \subseteq L(G)$ and whether $L \cap L(G) = \varnothing$, are solvable.

Example 3.1 Assume we are given the grammar

$$G = (\{S, A, B\}, \{a, b, c, +, (,)\}, S, F)$$

where the rules in F are

$$
\begin{array}{llll}
S \rightarrow S + A & A \rightarrow AB & B \rightarrow (S) & B \rightarrow b \\
S \rightarrow A & A \rightarrow B & B \rightarrow a & B \rightarrow c
\end{array}
$$

This grammar generates the language containing simple arithmetic expressions with addition and multiplication (in the guise of catenation) as the only operations and a, b, and c as the only variables.

Let us reduce this grammar to Chomsky normal form. First we introduce three new nonterminals C, D, and E and take the rules

$$
\begin{array}{llll}
S \rightarrow SCA & A \rightarrow AB & B \rightarrow DSE & B \rightarrow b \\
S \rightarrow A & A \rightarrow B & B \rightarrow a & B \rightarrow c \\
C \rightarrow + & D \rightarrow (& E \rightarrow) &
\end{array}
$$

Next we introduce two more nonterminals X and Y to reduce the length of the right-hand sides to a maximum of two.

$$
\begin{array}{llll}
S \rightarrow SX & A \rightarrow AB & B \rightarrow DY & B \rightarrow a \\
X \rightarrow CA & A \rightarrow B & Y \rightarrow SE & B \rightarrow b \\
S \rightarrow A & D \rightarrow (& E \rightarrow) & B \rightarrow c \\
C \rightarrow + & & &
\end{array}
$$

In order to eliminate $S \rightarrow A$ and $A \rightarrow B$ we define the sets

$$
\begin{array}{lll}
U(S) = \{S\} & U(C) = \{C\} & U(X) = \{X\} \\
U(A) = \{S, A\} & U(D) = \{D\} & U(Y) = \{Y\} \\
U(B) = \{S, A, B\} & U(E) = \{E\} &
\end{array}
$$

Then the rules in group I will be these:

$$
\begin{array}{llll}
S \rightarrow a & S \rightarrow b & S \rightarrow c & C \rightarrow + \\
A \rightarrow a & A \rightarrow b & A \rightarrow c & D \rightarrow (\\
B \rightarrow a & B \rightarrow b & B \rightarrow c & E \rightarrow)
\end{array}
$$

The rules in group II will be the following:

$$S \rightarrow SX$$

$$A \rightarrow AB \qquad S \rightarrow AB$$

$$B \rightarrow DY \qquad A \rightarrow DY \qquad S \rightarrow DY$$

$$X \rightarrow CA$$

$$Y \rightarrow SE$$

where the basic rules are listed in the first column and associated rules appear in the same line.

This way we have obtained an equivalent grammar in Chomsky normal form. It should be noted, however, that the Chomsky normal form is not unique. For instance, the rule $S \rightarrow SCA$ above can be replaced by $S \rightarrow XA$ and $X \rightarrow SC$ which leads to a different normal form.

3.2 DERIVATION TREE

A derivation in a context-free grammar can be represented very nicely by a directed graph where each node is labeled by a symbol from $V_N \cup V_T$, and a bunch of directed edges starting from the same node represents the application of a rule. Given, for instance, the grammar

$$G = (\{S, A, B\}, \{a, b, c\}, S, F)$$

with the rules in F

$$S \rightarrow ABc \qquad B \rightarrow aAc$$

$$A \rightarrow aB \qquad B \rightarrow bc$$

$$A \rightarrow Bc$$

Consider the derivation

$$S \Rightarrow ABc \Rightarrow AaAcc \Rightarrow BcaAcc \Rightarrow BcaaBcc \Rightarrow bccaaBcc \Rightarrow bccaabccc$$

This derivation is represented by the graph in Figure 3.1.

A directed, connected, and cycle-free graph is called a tree if it has no pair of edges terminating in the same node. This means that each node may have several outgoing edges but at most one entering edge. The precise definition is the following.

Definition 3.3 Let N be a finite nonvoid set whose elements are called *nodes*. The set of *edges* E is a set of ordered pairs of nodes, $E \subseteq N \times N$, and for an edge $e = (n_1, n_2) \in E$, $s(e) = n_1$ and $t(e) = n_2$ are called its

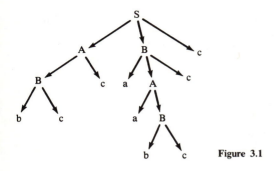

Figure 3.1

starting node and *terminating node*, respectively. A sequence of edges $e_0, e_1, \ldots, e_k$ is called a *directed path* of length k from $s(e_0)$ to $t(e_k)$ if $s(e_{i+1}) = t(e_i)$ for $i = 0, \ldots, k - 1$. The ordered pair (N, E) is called a *directed tree* if there is a node $r \in N$ such that

1) None of the terminating nodes of the edges in E is identical with r.
2) There is a unique directed path from r to every other node in N.

The node r is called the *root* of the tree and it follows immediately from the definition that every tree has a unique root. Each node of a tree is the root of a *subtree*. Extremal nodes having no outgoing edges are called *leaves*. The planar representation of a tree is not unique as we have not specified the order in which the edges starting from the same node are to be drawn for making a picture of the tree. In a derivation tree the order of those edges will be defined simply by the order of the symbols on the right-hand side of the corresponding rule (cf. Figure 3.1). This way we can associate a planar tree with every context-free derivation. The root of a derivation tree is labeled by S, every other node is labeled by a symbol from $V_N \cup V_T$. Naturally, a node labeled by some terminal symbol must be a leaf. If the leaves of a derivation tree are all labeled by terminal symbols, then it represents the derivation of a terminal word.

A derivation tree, however, does not always specify the order of the application of the rules. In Figure 3.1 we can see that $A \rightarrow Bc$ is applied before $B \rightarrow bc$ just below it, but we are free to choose the order of the application of $B \rightarrow aAc$ and the $A \rightarrow Bc$ to its left. This freedom of choice can be excluded by the requirement that in each step of the derivation the leftmost nonterminal symbol should always be replaced. It is clear that there is a unique *leftmost derivation* to every derivation tree. Two derivations are essentially the same if they differ only in the order of the application of the rules which means that they have the same derivation tree. Leftmost derivations can thus be considered as standard order derivations. Rightmost derivations have analogous properties and they will also be used later.

Derivation trees will be used to show a number of interesting properties of context-free languages.

Theorem 3.4 For any context-free grammar it is decidable whether it generates the empty language $\varnothing$.

PROOF It is enough to consider λ-free grammars only. Let us be given a grammar $G = (V_N, V_T, S, F)$ and let n denote the number of symbols in V_N. Assume further that the language $L(G)$ is not empty, i.e., there is a derivation $S \overset{*}{\underset{G}{\Rightarrow}} P$ where $P \in V_T^*$. Consider the tree associated with this derivation. If the longest path of this tree is longer than n then we can find some $P' \in L(G)$ having a derivation tree whose longest path is not longer than n.

Indeed, let the tree of the derivation $S \overset{*}{\underset{G}{\Rightarrow}} P$ contain a path of length $n + 1$. (See, e.g., the path $S \to B \to A \to B \to b$ in Figure 3.1.) Then we have $n + 1$ nodes labeled by nonterminal symbols along this path. But then, at least two nodes are labeled by the same symbol. (B is such a symbol in our example.) The first of these two nodes is nearer to the root S than the other. Replacing the subtree belonging to the former node by the subtree belonging to the latter we get a new derivation tree where the length of the given path is made definitely shorter and the leaves are still labeled by terminal symbols only (see Figure 3.2). By repeating this process we can cut back the length of every path which is longer than n. Thus, if the language contains any word at all then it must contain one with a derivation tree whose longest path is not longer than n.

In order to decide whether $L(G)$ is empty or not we may, therefore, restrict our search to derivation trees of limited size. A systematic search can be done by the following procedure:

STEP 1) Take the root S as the only element of the set (forest) T.

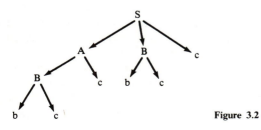

Figure 3.2

STEP 2) If there is a derivation tree in the set T which has only terminal symbols as labels of its leaves, then $L(G)$ is not empty and the procedure is finished.

STEP 3) Select each tree from T and extend it in every possible way by the application of a rule to one of its leaves labeled by a nonterminal symbol. Include each extended tree in the set T unless its longest path happens to be longer than n or it has already been included in T.

STEP 4) If during the last execution of step 3 no new tree has been added to T then $L(G)$ is empty and the procedure is finished. Otherwise it will be repeated from step 2.

Clearly, the number of derivation trees with at most n levels is finite for every grammar, so the above procedure will definitely stop after a finite number of steps. On the other hand, no derivation tree of this kind has been left out of consideration, and this completes the proof.

The decision of the emptyness of the language generated by a context-free grammar is also related to the question of redundancy of such grammars. A context-free grammar is called redundant if it contains unnecessary (useless) nonterminal symbols. A nonterminal symbol, i.e., a variable of a context-free grammar, can be useless in two different ways.

1) No terminal word is derivable from it.
2) It does not occur in any word derivable from the initial symbol of the grammar.

In the first case the variable is called *inactive* or dead, in the second case it is *unreachable*.

In order to decide whether or not a variable is inactive we can simply use that variable in place of the initial symbol of the grammar and see if it generates the empty language. That is, for any context-free grammar $G = (V_N, V_T, S, F)$, the variable $A \in V_N$ is inactive iff the language generated by

$$G_A = (V_N, V_T, A, F)$$

is empty.

The unreachability of A can be decided in the following way. Remove from F all rules which have A on their left-hand sides and denote the remaining rules by F_1. Then construct the grammar

$$G_A^\lambda = ((V_N - \{A\}) \cup V_T, \{A\}, S, F_1 \cup \{X \to \lambda \mid X \in (V_N - \{A\}) \cup V_T\})$$

Now, if A is unreachable, then $L(G_A^\lambda)$ is the language $\{\lambda\}$. (Otherwise it must contain a word with at least one occurrence of A.) Using the corollary of

Theorem 3.1 we can find a context-free grammar G' such that

$$L(G') = L(G_A^\lambda) - \{\lambda\}$$

and thus we have only to decide whether or not $L(G')$ is empty.

Inactive or unreachable variables, together with the rules in which they occur, can be eliminated from any context-free grammar without changing the language generated by it. So we can find an equivalent nonredundant context-free grammar to every context-free grammar. But the elimination of inactive or unreachable variables can be done more efficiently by using a direct approach instead of Theorem 3.4. We can more easily find the *active* and *reachable* variables than the inactive and the unreachable ones.

Definition 3.4 A variable of a context-free grammar is *active* if at least one terminal word is derivable from it and it is *reachable* if it occurs in at least one word derivable from the initial symbol of the grammar. A context-free grammar is *nonredundant* or *reduced* if each variable in it is both active and reachable.

Theorem 3.5 To every context-free grammar we can find an equivalent nonredundant grammar.

PROOF As we have seen above, this theorem follows from Theorem 3.4. But we can also give a direct proof this way. For the given grammar $G = (V_N, V_T, S, F)$ we define the following sets of variables

$$A_1 = \{X | X \to P \in F \text{ for some } P \in V_T^*\}$$

$$A_{i+1} = A_i \cup \{X | X \to W \in F \text{ for some } W \in (V_T \cup A_i)^*\}$$

Clearly, the sets A_i form a nondecreasing sequence

$$A_1 \subseteq A_2 \subseteq \cdots \subseteq A_i \subseteq A_{i+1} \subseteq \cdots$$

and they are all contained in V_N. Therefore, we have an index k such that $A_k = A_{k+1}$. For such a k $A_k = A_{k+j}$ is valid also for any $j \geqslant k$. Now, it is easy to see that A_k is precisely the set of the active variables of the grammar.

Similarly, we can define the set of reachable variables as follows:

$$R_1 = \{S\}$$

$$R_{i+1} = R_i \cup \{Y | X \to UYW \in F \text{ for some } X \in R_i \text{ and } U, W \in (V_N \cup V_T)^*\}$$

Here the limit R_m of the sequence

$$R_1 \subseteq R_2 \subseteq \cdots \subseteq R_i \subseteq R_{i+1} \subseteq \cdots$$

is exactly the set of all reachable variables. The computation of A_k and R_m takes usually a small number of steps. After that we have to eliminate all variables which do not belong in $A_k \cap R_m$ together with all rules in which they occur. The same process should be repeated with the resulting grammar until we get a nonredundant grammar.

The construction in the above proof can also be used to decide the emptyness problem. Namely, the language $L(G)$ is empty iff the initial symbol S is inactive.

Example 3.2 Find a reduced grammar that is equivalent to the context-free grammar

$$G = (\{S, A, B, C\}, \{a, b, c\}, S, F)$$

where F contains these rules

$$
\begin{array}{ll}
S \rightarrow aB & B \rightarrow aSB \\
S \rightarrow bC & B \rightarrow bBC \\
A \rightarrow BAc & C \rightarrow SBc \\
A \rightarrow bSC & C \rightarrow aBC \\
A \rightarrow a & C \rightarrow ac
\end{array}
$$

First we find the active variables.

$$
\begin{aligned}
A_1 &= \{A, C\} \\
A_2 &= \{A, C, S\} \\
A_3 &= \{A, C, S\}
\end{aligned}
$$

Therefore, the variable B is inactive. Next we find the reachable variables.

$$
\begin{aligned}
R_1 &= \{S\} \\
R_2 &= \{S, B, C\} \\
R_3 &= \{S, B, C\}
\end{aligned}
$$

Hence, the variable A is unreachable and thus, the reduced grammar is

$$G' = (\{S, C\}, \{a, b, c\}, S, \{S \rightarrow bC, C \rightarrow ac\})$$

The language generated by this grammar consists of a single word *bac*.

In our next theorem we shall use the derivation tree to establish a fundamental property of context-free languages. This theorem, also called the *pumping lemma*, was first shown by Bar-Hillel et al. [1961] and later refined by Ogden [1968] and by others. Here we present its original, i.e., nonrefined version.

Theorem 3.6 For every context-free language L we can give two natural numbers, p and q, such that each word in L that is longer than p has the form $UXWYZ$ where $|XWY| \leqslant q$, $XY \neq \lambda$, and UX^iWY^iZ is in L for all $i \geqslant 0$ (U, X, W, Y and Z are in V_T^*).

PROOF Only λ-free languages which can be generated by a grammar in Chomsky normal form are to be considered. Let n be the number of the elements of V_N and choose $p = 2^n$, $q = 2^{n+1}$. If $|P| > p$ for some $P \in L$, then the longest path in the derivation tree of P is longer than n. (It follows from the Chomsky normal form that no more than two edges start from each node. But a binary tree where the longest path is of length k can have at most 2^k leaves.) Consider the final $n + 1$ edges of the longest path. There is some variable $A \in V_N$ which occurs at least twice along this final section of the path. The subtree belonging to the first occurrence of A (nearer to the root S) represents a derivation $A \overset{*}{\Rightarrow} Q$ for some $Q \in V_T^*$. Similarly, we have some $W \in V_T^*$ and $A \overset{*}{\Rightarrow} W$ from the second occurrence of A. Moreover, there are two words X, Y in V_T^* such that $Q = XWY$. In fact, we obtain $A \overset{*}{\Rightarrow} XAY$ from the relative positions of the two occurrences of A. At the same time we have also some U and Z in V^* such that

$$S \overset{*}{\Rightarrow} UAZ \overset{*}{\Rightarrow} UXAYZ \overset{*}{\Rightarrow} UXWYZ = UQZ = P$$

The positions of the given occurrences of A in the tree imply that $|Q| \leqslant 2^{n+1}$. Further, the derivation $A \overset{*}{\Rightarrow} XAY$ involves the application of at least one $B \to CD$ type rule which means that $XY \neq \lambda$. Finally, we can see that the above derivations imply $S \overset{*}{\Rightarrow} UWZ$ as well as $S \overset{*}{\Rightarrow} UX^iWY^iZ$ for $i \geqslant 1$, which completes the proof.

Corollary 1 There are non-context-free phrase structure languages.

PROOF As we have seen in Example 1.2, the language

$$\{a^nb^nc^n \,|\, n = 1, 2, \ldots\}$$

is generated by a type 0 grammar. If it could be generated also by some type 2 grammar, then we would have some words U, X, W, Y, Z such that for every i UX^iWY^iZ is of the form $a^nb^nc^n$. But this is impossible since neither X nor Y may contain different letters so the number of the occurrences of the third letter cannot depend on i.

Corollary 2 It is decidable whether a context-free grammar generates an infinite language.

PROOF Only λ-free grammars are to be considered. We will show that $L(G)$ is infinite iff it contains some word P such that

$$p < |P| \leqslant p + q$$

where p and q are defined as in Theorem 3.6.

The if part of the assertion follows immediately from that theorem. To show the only if part we assume that L is infinite. Then it certainly contains a word P with $p < |P|$ so we have only to show that it contains one with $p < |P| \leqslant p + q$ as well. Assume, therefore, that for all $P \in L$ the condition $p < |P|$ implies $p + q < |P|$. But if $p < |P|$ then P has the form $UXWYZ$ with $UWZ \in L$ where $|UWZ| < |P|$ since $XY \neq \lambda$. If $p < |UWZ|$ then the argument can be repeated until we get some word $P' = U'X'W'Y'Z'$ with $p < |P'|$ and $|U'W'Z'| \leqslant p$. Now using the condition $|X'W'Y'| \leqslant q$ of the previous theorem we get $p < |U'X'W'Y'Z'| \leqslant p + q$ contrary to the indirect assumption.

The upper bound $p + q$ allows us to restrict our search for $P \in L$ to derivation trees of limited size (as we did it in Theorem 3.4), and this completes the proof.

REMARK The key step in the proofs of Theorems 3.4 and 3.6 is to find a variable A with the property that there is a derivation $A \overset{+}{\Rightarrow} XAY$ that involves more than zero steps for some $X, Y \in V_T^*$. Such a variable is called recursive and it allows for derivations with the form

$$A \overset{+}{\Rightarrow} X^i A Y^i \qquad (i \geqslant 1)$$

Thus, Theorem 3.6 can be interpreted as saying that every context-free grammar generating a sufficiently large word must have a recursive variable and, therefore, generates an infinite language.

3.3 LINEAR GRAMMARS AND REGULAR LANGUAGES

Definition 3.5 A context-free grammar is called *linear* if each rewriting rule has either of the two forms

1) $A \to P$ $A \in V_N$ $P \in V_T^*$
2) $A \to Q_1 B Q_2$ $A, B \in V_N$ and $Q_1, Q_2 \in V_T^*$

It is called, further, *left*-linear (*right*-linear) if Q_1 (Q_2) is λ in every rule of the second form.

According to this definition, right-linear grammars are exactly the same as type 3 (regular) grammars. It is easy to see that left-linear and right-linear grammars have the same generating power. (For an example, take the grammar of Example 2.1 and find an equivalent left-linear grammar.)

Theorem 3.7 Every left-linear grammar generates a type 3 language.

PROOF Assume that $G = (V_N, V_T, S, F)$ is a left-linear grammar and let $V_N = \{S, A_1, \ldots, A_n\}$. Without loss of generality we can assume that S does not occur on the right-hand sides of the rules. (Otherwise we can introduce S_0 and a new rule $S_0 \to S$.) We construct a right-linear grammar $G' = (V_N, V_T, S, F')$ with the production set F' as follows:

1) $S \to P \in F'$ iff $S \to P \in F$ and $P \in V_T^*$
2) $S \to PA_k \in F'$ iff $A_k \to P \in F$ and $P \in V_T^*$
3) $A_j \to PA_k \in F'$ iff $A_k \to A_j P \in F$ and $P \in V_T^*$
4) $A_j \to P \in F'$ iff $S \to A_j P \in F$ and $P \in V_T^*$

We can show that $L(G') = L(G)$ in the following way.

Let P be some word in $L(G)$. If $S \to P \in F$ then $S \to P \in F'$ and thus, $P \in L(G')$. Otherwise there is a derivation in G of the form

$$S \Rightarrow A_{i_1} P_1 \Rightarrow \cdots \Rightarrow A_{i_{m-1}} P_{m-1} \cdots P_1 \Rightarrow P_m \cdots P_1 = P$$

to which we can find a derivation in G' of the form

$$S \Rightarrow P_m A_{i_{m-1}} \Rightarrow P_m P_{m-1} A_{i_{m-2}} \Rightarrow \cdots \Rightarrow P_m \cdots P_2 A_{i_1} \Rightarrow P_m \cdots P_1$$

which implies $P \in L(G')$. So we have established the inclusion $L(G) \subseteq L(G')$. The reverse inclusion follows easily from the symmetry and this completes the proof.

Corollary $\mathcal{L}_3$ is closed under mirror image and every type 3 language is generated by a left-linear grammar.

PROOF Clearly, the mirror image of a right-linear grammar is left-linear. Thus, if $L \in \mathcal{L}_3$ then L^{-1} is generated by some left-linear grammar and $L^{-1} \in \mathcal{L}_3$ by Theorem 3.7. But then $(L^{-1})^{-1}$ is also generated by a left-linear grammar which proves also the second part of the corollary.

Now the question arises whether all linear grammars generate type 3 languages. The answer is in the negative which we shall prove using the following languages.

Example 3.3 Both of these languages

$$L_1 = \{a^n b^n c^k | n \geqslant 1, k \geqslant 1\}$$

and

$$L_2 = \{a^k b^n c^n | n \geqslant 1, k \geqslant 1\}$$

can be generated by linear grammars. The first is generated by

$$G = (\{S, A\}, \{a, b, c\}, S, F)$$

where the rules in F are

$$
\begin{array}{ll}
S \rightarrow Sc & A \rightarrow ab \\
S \rightarrow Ac & A \rightarrow aAb
\end{array}
$$

Clearly, L_2 is generated by similar rules.

It should be observed that the set-theoretical intersection of these languages is

$$L_1 \cap L_2 = \{a^n b^n c^n | n \geqslant 1\}$$

which is not context-free as we have seen before (corollary 1 of Theorem 3.6). On the other hand, it can be shown that the intersection of a regular and a context-free language is context-free which means that neither L_1 nor L_2 can be regular (see Theorem 3.9 below).

The theorem on the intersection of type 3 and type 2 languages can also be interpreted as a closure property of the class $\mathcal{L}_2$ under intersection with type 3 languages. In the proof of this theorem we shall use the following normal form theorem of type 3 grammars which is also interesting on its own.

Theorem 3.8 Every type 3 language can be generated by a grammar having the following two types of rules.

1) $X \rightarrow aY \quad X, Y \in V_N \quad a \in V_T$
2) $X \rightarrow \lambda \quad X \in V_N$

PROOF A type 3 grammar $G = (V_N, V_T, S, F)$ has two kinds of rules:

1) $A \rightarrow PB \quad A, B \in V_N \quad P \in V_T^*$
2) $A \rightarrow P \quad A \in V_N \quad P \in V_T^*$

Now, if $|P| > 1$, say, $P = a_1 a_2 \cdots a_n$ with $n \geqslant 2$ then the rule

$$A \rightarrow a_1 a_2 \cdots a_n B \quad (n \geqslant 2)$$

can be replaced by the set of rules

$$\{A \rightarrow a_1 Z_1, Z_1 \rightarrow a_2 Z_2, \ldots, Z_{n-1} \rightarrow a_n B\}$$

where $Z_1, \ldots, Z_{n-1}$ are newly introduced nonterminal symbols. Similarly, a rule of the form

$$A \rightarrow a_1 a_2 \cdots a_m \quad (m \geqslant 1)$$

can be replaced by the set of rules

$$\{A \rightarrow a_1 Y_1, Y_1 \rightarrow a_2 Y_2, \ldots, Y_{m-1} \rightarrow a_m Y_m, Y_m \rightarrow \lambda\}$$

where $Y_1, \ldots, Y_m$ are new nonterminals.

This way we can get an equivalent grammar with three types of rules:

1) $X \to aY$
2) $X \to Y$
3) $X \to \lambda$

where X and Y denote arbitrary nonterminal symbols and a denotes a terminal symbol. The chain rules of the form $X \to Y$ can now be eliminated in the same manner as in the proof of Theorem 3.2. Namely, assume that the set of rules has only the above three types of rules. More precisely, let $G = (V'_N, V_T, S, F_1)$ be already in that preliminary normal form. Let us define the sets $U(X)$ for all X in V'_N in the same way as given in the proof of Theorem 3.2. Then we define the set F' as follows.

1) $X \to aY \in F'$ iff there is some Z in V'_N such that $X \in U(Z)$ and $Z \to aY \in F_1$.
2) $X \to \lambda \in F'$ iff there is some Z in V'_N such that $X \in U(Z)$ and $Z \to \lambda \in F_1$.

Clearly, the grammar $G' = (V'_N, V_T, S, F')$ is equivalent to G and is in the required normal form which completes the proof. Now we can prove the above-mentioned closure property.

Theorem 3.9 The class $\mathcal{L}_2$ is closed under intersection with type 3 languages.

PROOF Let L be in $\mathcal{L}_2$ and L' be in $\mathcal{L}_3$. Then we have to show that $L \cap L'$ is in $\mathcal{L}_2$.

Assume first that $\lambda \notin L$. Then L can be generated by some grammar $G = (V_N, V_T, S, F)$ being in Chomsky normal form. On the other hand, L' can be generated by $G' = (V'_N, V'_T, S', F')$ being in the normal form established by Theorem 3.8. Let $\{X_1, \ldots, X_k\}$ be the set of all variables in V'_N for which

$$X_i \to \lambda \text{ is in } F' \qquad (i = 1, \ldots, k)$$

Consider now the context-free grammars

$$G_i = \left(V'_N \times (V_N \cup V_T) \times V'_N, V_T \cup V'_T, [S', S, X_i], F'' \right)$$

for $i = 1, \ldots, k$ where F'' is defined as follows.

1) $[X, A, Y] \to [X, B, Z][Z, C, Y] \in F''$ for all $X, Y, Z \in V'_N$ iff $A \to BC \in F$
2) $[X, A, Y] \to [X, a, Y] \in F''$ for all $X, Y \in V'_N$ iff $A \to a \in F$
3) $[X, a, Y] \to a \in F''$ iff $X \to aY \in F'$

Note that the nonterminals of G_i are ordered triples whose first and last components belong to V'_N while their second components belong to $V_N \cup V_T$. We shall prove that

$$L \cap L' = \bigcup_{i=1}^{k} L(G_i)$$

which combined with Theorem 2.2 will give the result.

A word $P = a_1, a_2, \ldots, a_n \in V_T^*$ is derivable in G if and only if for every i and for every sequence of nonterminals $Z_1, \ldots, Z_{n-1}$ in V'_N there is a derivation in G_i with the form

$$[S', S, X_i] \underset{G_i}{\overset{*}{\Rightarrow}} [S', a_1, Z_1][Z_1, a_2, Z_2] \cdots [Z_{n-1}, a_n, X_i]$$

Further, P is derivable in G' if and only if there is a sequence of nonterminals $Z_1, \ldots, Z_{n-1}$ in V'_N and some $X_i \in V'_N$ with $X_i \to \lambda \in F'$ such that

$$S' \underset{G'}{\Rightarrow} a_1 Z_1 \underset{G'}{\Rightarrow} a_1 a_2 Z_2 \underset{G'}{\Rightarrow} \cdots \underset{G'}{\Rightarrow} a_1 a_2 \cdots a_n X_i \underset{G'}{\Rightarrow} a_1 a_2 \cdots a_n$$

Hence, by the construction of G_i we conclude that P is derivable in G' iff there are some $Z_1, \ldots, Z_{n-1}$ in V'_N and there is a derivation in G_i such that

$$[S', a_1, Z_1][Z_1, a_2, Z_2] \cdots [Z_{n-1}, a_n, X_i] \underset{G_i}{\overset{*}{\Rightarrow}} a_1 a_2 \cdots a_n$$

This means that $P \in L \cap L'$ iff $P \in L(G_i)$ for some i.

In case $\lambda \in L$ we can construct a context-free grammar to generate $(L - \langle \lambda \rangle) \cap L'$ and supplement it by the rules $S_0 \to \lambda$ and $S_0 \to S$ iff $\lambda \in L'$.

On the basis of Example 3.3 we can formulate two important corollaries.

Corollary 1 The class $\mathcal{L}_3$ is properly included in $\mathcal{L}_2$.

Corollary 2 The class $\mathcal{L}_3$ is properly included in the class of linear languages.

This way we have settled the question of the relation between $\mathcal{L}_3$ and $\mathcal{L}_2$ but we do not know what makes a linear grammar generate a nonregular language. More insight into the problem can be gained by considering the self-embedding property.

Definition 3.6 A context-free grammar $G = (V_N, V_T, S, F)$ is called self-embedding if it has a variable $A \in V_N$ with $A \overset{*}{\Rightarrow} XAY$ for some $X, Y \in (V_N \cup V_T)^+$.

Theorem 3.10 If a context-free grammar G is not self-embedding, then $L(G) \in \mathcal{L}_3$.

PROOF In virtue of Theorem 3.5 we can assume that G is reduced. Two cases will be distinguished.

CASE 1 For each $A \in V_N$ there is a derivation $A \overset{*}{\Rightarrow} USZ$ with $U, Z \in (V_N \cup V_T)^*$.

Each rule which contains at least one variable on the right-hand side has either of these forms:

1) $A \to XBY$
2) $A \to XB$
3) $A \to BY$
4) $A \to B$

where $A, B \in V_N$, and $X, Y \in (V_N \cup V_T)^+$. If there is a rule in F with the form 1 then there exists a derivation of form

$$A \Rightarrow XBY \overset{*}{\Rightarrow} XUSZY \overset{*}{\Rightarrow} XUPAQZY$$

which is in contradiction with the non-self-embedding of G. If there are rules of form 2 and also of form 3, say $A \to XB$ and $C \to DY$, then the same contradiction arises from the derivation

$$A \Rightarrow XB \overset{*}{\Rightarrow} XU_1 SZ_1 \overset{*}{\Rightarrow} XU_1 P_1 CQ_1 Z_1 \Rightarrow XU_1 P_1 DYQ_1 Z_1 \overset{*}{\Rightarrow}$$

$$XU_1 P_1 U_2 SZ_2 YQ_1 Z_1 \overset{*}{\Rightarrow} XU_1 P_1 U_2 P_2 AQ_2 Z_2 YQ_1 Z_1$$

The situation is similar if we have a form 2 rule with $X \notin V_T^*$ or a form 3 rule with $Y \notin V_T^*$. Therefore, the rules in F must have forms either $A \to XB$ and $A \to X$ with $A, B \in V_N$, $X \in V_T^*$ or else $A \to BY$ and $A \to Y$ with $A, B \in V_N$, $Y \in V_T^*$ which implies that $L(G) \in \mathcal{L}_3$.

CASE 2 There is at least one variable $A_1 \in V_N$ such that $A_1 \overset{*}{\Rightarrow} USZ$ does not hold for any $U, Z \in (V_N \cup V_T)^*$.

In this case we use induction on the number of the variables in V_N. If V_n has only one member then this case cannot occur since $S \overset{*}{\Rightarrow} S$ always holds. Assume now that the theorem is true for every grammar with at most n variables and let G have $n + 1$ of them. Then take the grammar

$$G_1 = (V_N - \{S\}, V_T, A_1, F_1)$$

where F_1 is obtained from F by leaving out all rules containing S. Take also the grammar

$$G_2 = (V_N - \{A_1\}, V_T \cup \{A_1\}, S, F_2)$$

where F_2 is obtained from F by leaving out the rules with A_1 on the left-hand side. Both G_1 and G_2 are non-self-embedding as $F_1 \subseteq F$ and $F_2 \subseteq F$. Hence, by the induction hypothesis both $L(G_1)$ and $L(G_2)$ are type 3 languages. At this point, unfortunately, we have to make a forward reference to Definition 3.9 in Section 3.5. Namely, the language $L(G)$ can be obtained by substituting $L(G_1)$ for A_1 and $\{a\}$ for all $a \in V_T$ in $L(G_2)$. Since the class $\mathcal{L}_3$ is closed under substitution, we conclude that $L(G)$ is also type 3, which completes the proof.

3.4 GREIBACH NORMAL FORM

As we have seen in Section 3.2 every derivation in a context-free grammar can be made left-to-right by choosing to rewrite the leftmost variable in each step of the derivation. Intuitively, this is due to the fact that each nonterminal is the root of a subtree in the derivation tree and each subtree is developed independently from the other. More formally we can show this by considering an arbitrary derivation

$$S \Rightarrow P_1 \Rightarrow P_2 \Rightarrow \cdots \Rightarrow P_m = P$$

where $P_i = Q_i A R_i$ and $P_{i+1} = Q_i X R_i$ such that $Q_i, R_i \in (V_N \cup V_T)^*$ and $A \rightarrow X \in F$. If this derivation is leftmost, then $Q_i \in V_T^*$ for $i = 1, \ldots, m - 1$. In the opposite case let j be the smallest index for which $P_j = Q_j A R_j$, $P_{j+1} = Q_j X R_j$, $A \rightarrow X \in F$ and $Q_j \notin V_T^*$. Then $Q_j = T_j B U_j$ with $T_j \in V_T^*$, $B \in V_N$, and $U_j \in (V_N \cup V_T)^*$. But in the given derivation there must be a step where some rule $B \rightarrow Y$ is applied to the given occurrence of B. By changing the order of the application of the rules we can achieve that $P_j = T_j B U_j A R_j$ and $P_{j+1} = T_j Y U_j A R_j$ will be the case. Repeating this procedure for the rest of the derivation we get a leftmost derivation with the same number of steps.

The idea of leftmost derivation is thus related to the attempt of increasing the number of terminal symbols at the beginning of the word in each step of the derivation. An increasing in the strict sense is obviously impossible with a rule of form $A \rightarrow BX$ where $B \in V_N$. Interestingly enough, every λ-free grammar can be transformed into a normal form where the number of the terminal symbols at the beginning of the word will be strictly increased in each step of a left-most derivation.

Definition 3.7 A context-free grammar is said to be in *Greibach normal form* if each rule is of form $A \rightarrow aX$ with $A \in V_N$, $a \in V_T$, and $X \in V_N^*$.

Note that the grammar given in Example 1.1 is in Greibach normal form.

Theorem 3.11 To every λ-free context-free grammar we can give an equivalent grammar in Greibach normal form.

PROOF Let $G = (V_N, V_T, S, F)$ be a λ-free context-free grammar with $V_N = \{A_1, \ldots, A_n\}$. According to Theorem 2.1 (see the remark at the end of the proof) we can assume that each rule has either of the two forms

1) $A_i \to a \qquad a \in V_T$
2) $A_i \to A_j X \qquad X \in V_N^*$

First we want to show how $i \neq j$ can be achieved for any rule in group (2) at the cost of introducing some other rules. Let

$$\{A_k \to A_k X_1, \ldots, A_k \to A_k X_r\} \qquad (X_i \in V_N^+ \text{ for } i = 1, \ldots, r)$$

be the set of the "unwanted" rules for some A_k and let

$$\{A_k \to Y_1, \ldots, A_k \to Y_s\}$$

be the set of all the other rules with A_k on the left-hand side. Replace the set of "unwanted" rules by the set of new rules

$$A_k \to Y_j Z_k \qquad \text{for } j = 1, \ldots, s$$
$$Z_k \to X_i \qquad \text{for } i = 1, \ldots, r$$
$$Z_k \to X_i Z_k \qquad \text{for } i = 1, \ldots, r$$

where $Z_k \notin V_N$ is a newly introduced variable. The new grammar generates the same language as the original one because to each derivation in the original grammar of the form

$$A_k \Rightarrow A_k X_{i_1} \Rightarrow \cdots \Rightarrow A_k X_{i_m} \cdots X_{i_1} \Rightarrow Y_j X_{i_m} \cdots X_{i_1}$$

there is a derivation with the new rules

$$A_k \Rightarrow Y_j Z_k \Rightarrow Y_j X_{i_m} Z_k \Rightarrow \cdots \Rightarrow Y_j X_{i_m} \cdots X_{i_2} Z_k \Rightarrow Y_j X_{i_m} \cdots X_{i_1}$$

and conversely, to each derivation with the new rules there is a derivation in the original grammar.

Returning to our original grammar we divide its rules into two groups:

1) $A_i \to aX$
2) $A_i \to A_j X$

with $X \in V_N^*$ for each group. Our first aim is to achieve $i < j$ for each rule in group 2.

Let k be the greatest integer such that $k \leqslant n + 1$ and $i < j$ holds for every rule $A_i \to A_j X$ with $i < k$. If $k = n + 1$ then we are done; otherwise there must be a rule $A_k \to A_j X_0$ with $k \geqslant j$. If $k > j$ with this rule then it will be replaced by the set of rules of form $A_k \to Y X_0$ for all Y such that $A_j \to Y \in F$. (Such a replacement is nothing but a simple substitution for

A_j.) Each Y here begins with either a terminal symbol or some A_m where $j < m$ and the generated language does not change by this replacement. By repeating this process (if necessary) we can eliminate every rule of the form $A_k \rightarrow A_j X$ with $k > j$ so that only the case $k = j$ might remain. Then we eliminate also the rules (if any) of the form $A_k \rightarrow A_k X$ using the method described before. Thus, $i < j$ will hold for every rule $A_i \rightarrow A_j X$ with $i \leqslant k$. Therefore, we compute the new value for k and repeat the whole process until we get $k = n + 1$. At the end we shall have three kinds of rules:

1) $A_i \rightarrow aX$ $a \in V_T$
2) $A_i \rightarrow A_j X$ $i < j$
3) $Z_i \rightarrow X$

where $X \in (V_N \cup \langle Z_1, \ldots, Z_n \rangle)^*$. With A_n on the left-hand side we can have only form 1 rules. By substituting the right-hand sides of these rules for A_n in every $A_i \rightarrow A_n X$ we get form 1 rules. Then we substitute for A_{n-1}, A_{n-2}, and so on, until all form 2 rules are eliminated. The construction of form 3 rules implies that the first symbol on the right-hand side of such a rule may only be one of the original variables A_i. The latter can be eliminated by substitution as in the case of form 2 rules and this completes the proof.

REMARK The derivation of a terminal word P in a grammar that is in Greibach normal form has always $|P|$ steps. The Greibach normal form shows some similarity to a type 3 grammar but the right-hand sides of the rules in Greibach normal form may well contain more than one variable and this makes the difference.

Example 3.4 Find an equivalent grammar in Greibach normal form to the grammar given in Example 3.1. We start with the rules

$$S \rightarrow SCA \qquad A \rightarrow AB \qquad B \rightarrow DSE \qquad B \rightarrow b$$
$$S \rightarrow A \qquad A \rightarrow B \qquad B \rightarrow a \qquad B \rightarrow c$$
$$C \rightarrow + \qquad D \rightarrow (\qquad E \rightarrow)$$

First arrange the variables in some sequence. It is advisable to choose the order of the variables in such a way as to minimize the work required to eliminate inconvenient rules. Let us take, therefore, the order S, A, B, C, D, E. Then the first rule to eliminate is $S \rightarrow SCA$. So we get

$$S \rightarrow AZ \qquad Z \rightarrow CA \qquad Z \rightarrow CAZ \qquad A \rightarrow AB$$
$$B \rightarrow DSE \qquad B \rightarrow a \qquad B \rightarrow b \qquad B \rightarrow c$$
$$S \rightarrow A \qquad A \rightarrow B \qquad C \rightarrow + \qquad D \rightarrow ($$
$$E \rightarrow)$$

Next we eliminate $A \to AB$.

$$S \to AZ \qquad Z \to CA \qquad Z \to CAZ$$
$$A \to BU \qquad U \to B \qquad U \to BU$$
$$S \to A \qquad A \to B \qquad B \to DSE$$
$$B \to a \qquad B \to b \qquad B \to c$$
$$C \to + \qquad D \to (\qquad E \to)$$

Now we can substitute for D, C, and B at once.

$$S \to AZ \qquad Z \to +A \qquad Z \to +AZ$$
$$A \to aU \qquad U \to a \qquad U \to aU$$
$$A \to bU \qquad U \to b \qquad U \to bU$$
$$A \to cU \qquad U \to c \qquad U \to cU$$
$$A \to (SEU \qquad U \to (SE \qquad U \to (SEU$$
$$S \to A \qquad A \to a \qquad A \to b$$
$$A \to c \qquad A \to (SE \qquad B \to (SE$$
$$B \to a \qquad B \to b \qquad B \to c$$
$$C \to + \qquad D \to (\qquad E \to)$$

The last step is the substitution for A in $S \to AZ$ and $S \to A$, which gives

$$S \to aUZ \qquad S \to bUZ \qquad S \to cUZ \qquad S \to (SEUZ$$
$$S \to aZ \qquad S \to bZ \qquad S \to cZ \qquad S \to (SEZ$$
$$S \to aU \qquad S \to bU \qquad S \to cU \qquad S \to (SEU$$
$$S \to a \qquad S \to b \qquad S \to c \qquad S \to (SE$$
$$A \to aU \qquad A \to bU \qquad A \to cU \qquad A \to (SEU$$
$$A \to a \qquad A \to b \qquad A \to c \qquad A \to (SE$$
$$U \to aU \qquad U \to bU \qquad U \to cU \qquad U \to (SEU$$
$$U \to a \qquad U \to b \qquad U \to c \qquad U \to (SE$$
$$Z \to +A \qquad Z \to +AZ \qquad E \to)$$

The rules with B, C, or D on the left-hand side have been left out for these variables do not occur on the right-hand side of any rule. As can be seen in this example the Greibach normal form is rather detailed and a more compact form is often preferable.

3.5 REGULAR EXPRESSIONS

Besides right-linear (or left-linear) grammars, type 3 languages have another interesting characterization. We know that every finite language belongs to $\mathcal{L}_3$. (For each word P in L we can have a production $S \to P$.) Further, we know by Theorem 2.2 that $\mathcal{L}_3$ is closed under the regular operations which are the set union, the catenation, and the closure of iteration. Therefore, we can define a type 3 language by starting with a finite number of words and then applying regular operations to them. This method gives rise to the so-called *regular expressions* which are formally defined in the following way.

Definition 3.8 A regular expression over a finite alphabet V is defined inductively as follows:

1) λ (i.e., the empty string) is a regular expression.
2) a is a regular expression for every a in V.
3) If R is a regular expression over V, then so is $(R)^*$.
4) If Q and R are regular expressions over V then so are $(Q);(R)$ and $(Q)|(R)$.

The symbols *, ;, and | used in Definition 3.8 denote, respectively, the operations of iteration, catenation, and set union. As mentioned before, every regular expression defines—here we shall say *denotes*—some regular language. (λ denotes $\{\lambda\}$, a denotes $\{a\}$, $a|b$ denotes $\{a, b\}$, and so forth.) The interesting question is whether the converse is also true. To our great satisfaction the answer is yes, namely, every type 3 language can be described in form of a regular expression. Before proving this result let us see a few examples.

Example 3.5 Consider the alphabet $V = \{a, b\}$. Then each of the following is a regular expression denoting the language next to it. Many of the parentheses may be omitted if we define the priority of the operations. The usual order is *, ;, |, which makes regular expressions easier to write.

a^* is the same as $(a)^*$ and denotes $\{a\}^*$
$(a|b)^*$ is the same as $((a)|(b))^*$ and denotes $\{a, b\}^*$
$a^*; b$ is the same as $((a)^*);(b)$ and denotes $\{a\}^*\{b\}$
$b|a; b^*$ is the same as $(b)|((a);((b)^*))$ and denotes $\{b\} \cup \{a\}\{b\}^*$
$(a|b); a^*$ is the same as $((a)|(b));((a)^*)$ and denotes $\{a, b\}\{a\}^*$.

Note that the language $\{a, b\}\{a\}^*$ is the same as $\{a\}\{a\}^* \cup \{b\}\{a\}^*$. This means that

$$(a|b); a^* = a; a^*|b; a^*$$

is a valid equation between two regular expressions with respect to their meaning (denotation). There are many interesting equations between regular expressions which can be obtained by substitution for P, Q, and R in the following axioms which are valid equations for all regular expressions P, Q, and R.

$$P|(Q|R) = (P|Q)|R \qquad P;(Q;R) = (P;Q);R$$

$$P|Q = Q|P \qquad P;(Q|R) = P;Q|P;R$$

$$(P|Q);R = P;R|Q;R \qquad P^* = \lambda|P;P^*$$

$$\lambda;P = P;\lambda = P \qquad P^* = (\lambda|P)^*$$

Unfortunately, not every valid equation can be deduced from these axioms nor from any finite set of axioms by using substitution only. We need another inference rule, namely, a conditional equation of the form

$$\text{if } P = R|P;Q \quad \text{and} \quad \lambda \notin Q \quad \text{then } P = R;Q^*$$

(Observe the similarity of this rule to the solution of the mathematical equation $p = r + pq$ which is $p = r(1-q)^{-1}$ for $q \neq 1$. Further, if $|q| < 1$ then we have the infinite series expansion

$$(1-q)^{-1} = 1 + q + q^2 + \cdots$$

which corresponds to Q^* in our case.) For the sake of completeness we can add $\varnothing$ (denoting the empty language) to the set of regular expressions. In that case we do not need λ, because $\{\lambda\} = \varnothing^*$. So we can replace part 1 of Definition 3.8 by

1') $\varnothing$ is a regular expression (denoting the empty language).

Then we have to replace λ by $\varnothing^*$ in the above axiom system and add one more axiom

$$\varnothing;P = P;\varnothing = \varnothing$$

This axiom system and the two inference rules, namely, substitution and the above conditional equation, are sufficient for deducing every valid equation between regular expressions. Here we do not prove this completeness theorem. The interested reader is referred to the literature, in particular, Salomaa [1966] and Ginzburg [1968]. Next we prove our main theorem about regular expressions.

Theorem 3.12 Every regular expression denotes a type 3 language and, conversely, every type 3 language is denoted by a regular expression.

PROOF The first part of the theorem, as mentioned before, follows directly from Theorem 2.2.

For the second part, let $L = L(G)$ where $G = (V_N, V_T, S, F)$ is a right-linear grammar in the normal form of Theorem 3.8. Let $V_N =$

$\langle A_1, \ldots, A_n \rangle$ be the set of variables with $S = A_1$ being the initial symbol. (Each rule in F has form either $A_i \to aA_j$ with $a \in V_T$ or $A_i \to \lambda$.) We say that a derivation of the form $A_i \overset{*}{\Rightarrow} PA_j$ *involves* A_m iff A_m occurs in an intermediate string between A_i and PA_j in that derivation. (It should be emphasized that the derivation $A_i \overset{*}{\Rightarrow} PA_j$ does not involve A_i and A_j unless they occur in between as well.) The derivation $A_i \overset{*}{\Rightarrow} PA_j$ is said to be *k-restricted* iff $0 \leqslant m \leqslant k$ for every A_m which is involved in that derivation. Consider now the sets of words

$$E_{ij}^k = \left\{ P \in V_T^* \,|\, \text{there is a k-restricted derivation } A_i \overset{*}{\Rightarrow} PA_j \right\}$$

First we prove by induction on k that E_{ij}^k is denoted by a regular expression for every i, j, and k with $0 \leqslant i, j, k \leqslant n$.

Basis For $i \neq j$ the set E_{ij}^0 is either empty or consists of some letters of V_T. ($a \in E_{ij}^0$ iff $A_i \to aA_j \in F$.) For $i = j$ the set E_{ii}^0 contains λ and zero or more letters of V_T. In any case E_{ij}^0 is denoted by a regular expression.

Induction step Assume that for a fixed value of k with $0 < k \leqslant n$ each of the sets E_{ij}^{k-1} is denoted by a regular expression. It should be clear that for every i, j, and k we have

$$E_{ij}^k = E_{ij}^{k-1} | E_{ik}^{k-1} ; \left(E_{kk}^{k-1} \right)^* ; E_{kj}^{k-1}$$

since A_k may only be involved as expressed by this equation. Hence, by the induction hypothesis E_{ij}^k is also denoted by a regular expression.

Now, to finish our proof let I_λ denote the set of indices such that $i \in I_\lambda$ iff $A_i \to \lambda \in F$. Then obviously

$$L(G) = \bigcup_{i \in I_\lambda} E_{1i}^n$$

which is denoted by a regular expression, and this completes the proof.

The concept of regular expressions suggests another operation on languages called substitution.

Definition 3.9 Given a finite alphabet V, let V_a denote an alphabet and $s(a) \subseteq V_a^*$ denote a language for each $a \in V$. For each word $P = a_1 a_2 \cdots a_n \in V^*$ we define the substitution

$$s(P) = s(a_1)s(a_2) \cdots s(a_n)$$

as the catenation of the languages corresponding to the letters of P. This is extended to any $L \subseteq V^*$ by the formula

$$s(L) = \langle Q | Q \in s(P) \text{ for some } P \in L \rangle$$

By using regular expressions it is easy to see that the class $\mathcal{L}_3$ is closed under substitution. That is, the set of regular expressions is clearly closed under substitution of a regular expression for each of its letters. Substitution can be considered as the generalization of homomorphism given in Chapter 2, Definition 2.1.

EXERCISES

3.1 Eliminate the λ-rules from the context-free grammar

$$G = (\{S, A, B, C, D, E, Z\}, \{a, b, c, d, e, f, g\}, S, F)$$

with the rules in F as follows

$S \rightarrow aB$	$D \rightarrow eE$
$B \rightarrow \lambda$	$E \rightarrow \lambda$
$B \rightarrow bCd$	$E \rightarrow faZ$
$C \rightarrow D$	$Z \rightarrow \lambda$
$C \rightarrow CcD$	$Z \rightarrow ge$
$D \rightarrow a$	

3.2 Find a reduced context-free grammar equivalent to

$$G = (\{S, A, B, C, D\}, \{a, b, c\}, S, F)$$

where the rules in F are

$S \rightarrow aBc$	$B \rightarrow Abc$
$S \rightarrow AbC$	$B \rightarrow abC$
$S \rightarrow \lambda$	$B \rightarrow SBc$
$A \rightarrow aBC$	$C \rightarrow BSC$
$A \rightarrow ABc$	$C \rightarrow Abc$
$A \rightarrow AbS$	$D \rightarrow SaB$
$A \rightarrow SbC$	$D \rightarrow ab$

3.3 Give a context-free grammar that generates the language containing all finite decimal numbers, integers, and fractions, occasionally with the $-$ sign in front. (Leading zeroes are not permitted for integer parts.) A two-digit exponent of the form $E \pm d_1 d_2$ may also be attached, where d_1 and d_2 are arbitrary decimal digits. Find an equivalent grammar in Chomsky normal form.

3.4 Find an equivalent grammar in Greibach normal form to the grammar given in Example 3.3 and then transform it to Chomsky normal form.

3.5 Find a type 3 (i.e., right-linear) grammar that generates the mirror image of the language given in Example 2.1.

3.6 Prove that the language $\langle a^{n^2}|n = 1, 2, \dots \rangle$ is not context-free. *Hint:* Compare the length of subsequent words in the sequence obtained from the pumping lemma (Theorem 3.6).

3.7 Find a context-free grammar in Greibach normal form to generate the language $\langle PP^{-1}|P \in \langle a, b \rangle^* \rangle$.

3.8 Find a regular expression denoting the language described in Exercise 3.3.

3.9 Prove that the language $\langle a^{2i}b^{i+j}c^{2j}|i, j = 0, 1, \dots \rangle$ is not regular.

3.10 Consider the grammar $G = (\langle S, A, B \rangle, \langle a, b \rangle, S, F)$ where F contains the following rules:

$$S \rightarrow aB \qquad S \rightarrow bA$$
$$A \rightarrow aS \qquad A \rightarrow bAA \qquad A \rightarrow a$$
$$B \rightarrow bS \qquad B \rightarrow aBB \qquad B \rightarrow b$$

For the string *aaabbabbba* find two different derivation trees (and hence two different leftmost derivations).

3.11 Find a grammar in Chomsky normal form to generate the language $\langle a^n PP^{-1}b^n|n = 1, 2, \dots \rangle$.

3.12 Is the language $\langle a^m b^n|m \leqslant n \leqslant 2m \rangle$ context-free?

3.13 Find a type 3 grammar to generate the language denoted by the regular expression $a; (a|b)^*|b^*; a$.

3.14 Give a context-free grammar to generate the set of all regular expressions over the alphabet $\langle a, b \rangle$.

3.15 Without using regular expressions, prove that the class $\mathcal{L}_3$ is closed under substitution.

3.16 Show that $\mathcal{L}_2$ is not closed under complementation. *Hint:* Make use of set theory.

CHAPTER
FOUR

CONTEXT-SENSITIVE LANGUAGES

4.1 LENGTH-INCREASING GRAMMARS

In this section we will show that the family of context-free languages is properly contained in the family of context-sensitive ones. In other words, the generating power of context-sensitive grammars is effectively greater than that of the context-free ones. Apart from this fact, context-sensitive grammars can also be useful for dealing with context-free languages. We shall return to this point in Chapter 9 in connection with syntax analysis. Intuitively it should be clear that contextual information may be useful in processing any language. Here we shall prove only the basic properties of context-sensitive grammars.

Definition 4.1 A type 0 grammar is called *length-increasing* (more precisely, nondecreasing) if for all rules $P \to Q$ in F we have $|P| \leqslant |Q|$.

Clearly, every λ-free context-sensitive grammar is length-increasing. Further we show:

Theorem 4.1 Every length-increasing grammar generates a context-sensitive language.

PROOF Let $G = (V_N, V_T, S, F)$ be a length-increasing grammar. It can be assumed that terminal symbols occur only on the right-hand sides of the rules with form $A \to a$. Now let $P \to Q$ in F such that $|P| \geqslant 2$. Then it is of form

$$X_1 X_2 \cdots X_m \to Y_1 Y_2 \cdots Y_n \qquad (2 \leqslant m \leqslant n)$$

where X_i, Y_j are in V_N. This rule can be replaced by the set of the following context-sensitive rules:

$$X_1 X_2 \cdots X_m \to Z_1 X_2 \cdots X_m$$
$$Z_1 X_2 \cdots X_m \to Z_1 Z_2 \cdots X_m$$
$$\vdots$$
$$Z_1 \cdots Z_{m-1} X_m \to Z_1 \cdots Z_{m-1} Z_m Y_{m+1} \cdots Y_n$$
$$Z_1 \cdots Z_{m-1} Z_m Y_{m+1} \cdots Y_n \to Y_1 \cdots Z_{m-1} Z_m Y_{m+1} \cdots Y_n$$
$$\vdots$$
$$Y_1 \cdots Y_{m-1} Z_m Y_{m+1} \cdots Y_n \to Y_1 \cdots Y_{m-1} Y_m Y_{m+1} \cdots Y_n$$

where $Z_i \notin V_N$ $(i = 1, \ldots, m)$ are newly introduced variables. This way we can replace every length-increasing rule by context-sensitive ones without changing the generated language and this completes the proof.

Corollary $\mathcal{L}_2$ is properly included in $\mathcal{L}_1$.

PROOF The language $\{a^{n^2} | n = 1, 2, \ldots\}$ is generated by a length-increasing grammar (see Exercise 1.5); hence it is context-sensitive, but according to Exercise 3.6, it is not context-free. (The language $\{a^n b^n c^n | n = 1, 2, \ldots\}$ can be used just as well in this proof.)

The length-increasing property is thus equivalent to context-sensitivity with the sole exception of the rule $S \to \lambda$ that is needed only to derive the empty word λ. We may ask whether each type 0 language can be generated by some length-increasing grammar. This question turns out to be a very difficult one and we shall see only in Chapter 7 that the answer is negative.

Derivations in context-sensitive grammars are in general much more complicated than in the context-free case though they are—at least theoretically—all manageable due to the length-increasing property.

Theorem 4.2 For every context-sensitive grammar G and for an arbitrary word P, it is decidable whether or not P is in $L(G)$.

PROOF For $P = \lambda$, the answer is trivial. For $P \in V_T^+$, consider all finite sequences of words

$$S = P_0, P_1, P_2, \ldots, P_n = P$$

such that $P_i \in (V_N \cup V_T)^+$ and $|P_i| \leqslant |P_{i+1}|$ for $i = 0, 1, \ldots, n-1$ and $P_i \neq P_j$ if $i \neq j$. The total number of words Q in $(V_N \cup V_T)^+$ with $|Q| \leqslant |P|$ for any given P is finite; thus, the number of such sequences without repetitions is also finite. Hence, we can produce systematically all such sequences and check whether $P_i \overset{*}{\Rightarrow} P_{i+1}$ $(i = 0, 1, \ldots, n-1)$ holds for at least one of them. If so, then clearly $P \in L(G)$, otherwise P cannot be in $L(G)$. (The shortest derivation of P cannot have repetitions.)

This decision procedure is rather impractical. It generally takes a tremendous number of steps, so we must look for more efficient procedures which work faster, at least in some special cases. Such techniques are widely used in compiler construction for programming languages, but only for the context-free case.

For an arbitrary type 0 grammar, however, this procedure would not work because length-increasing and length-decreasing steps may follow each other in a derivation sequence so we cannot establish any fixed bound to the length of the shortest derivation of P. This shows the difficulty of the membership problem, i.e., the problem of deciding whether or not P is in $L(G)$, in case of type 0 grammars.

4.2 KURODA NORMAL FORM

Definition 4.2 A length-increasing grammar is said to be in Kuroda normal form if each of its rules has any of the following four forms:

1) $A \rightarrow a$
2) $A \rightarrow B$
3) $A \rightarrow BC$
4) $AB \rightarrow CD$

Theorem 4.3 For every length-increasing grammar, we can give an equivalent grammar in Kuroda normal form.

PROOF Again we can assume that terminal symbols occur only in $A \rightarrow a$ form rules. If $P \rightarrow Q \in F$ such that $|P| = 1$ and $|Q| > 2$ then we can replace it by form 3 rules in the same way as we have done it for the Chomsky normal form. If $|P| = 2$ and $|Q| = 2$ then $P \rightarrow Q$ is of form 4. So we have to deal only with the case $|P| \geqslant 2$ and $|Q| > 2$. A rule of this form

$$X_1 \cdots X_m \rightarrow Y_1 \cdots Y_n \qquad (2 \leqslant m \leqslant n)$$

with $X_i, Y_j \in V_N$ can be replaced by the set of rules

$$
\begin{array}{ll}
X_1 X_2 \rightarrow Y_1 Z_2 & Z_m \rightarrow Y_m Z_{m+1} \\
Z_2 X_3 \rightarrow Y_2 Z_3 & Z_{m+1} \rightarrow Y_{m+1} Z_{m+2} \\
\quad \vdots & \quad \vdots \\
Z_{m-1} X_m \rightarrow Y_{m-1} Z_m & Z_{n-1} \rightarrow Y_{n-1} Y_n
\end{array}
$$

where $Z_2, \ldots, Z_{n-1}$ are new variables. Performing such replacements as often as required we get an equivalent grammar in Kuroda normal form.

Corollary Every λ-free context-sensitive language can be generated by a grammar in Kuroda normal form.

REMARK Each $AB \to CD$ type rule in the Kuroda normal form can be replaced by four context-sensitive rules as follows:

$$AB \to AB' \qquad A'B' \to A'D$$
$$AB' \to A'B' \qquad A'D \to CD$$

where A' and B' are new variables. This means that Theorem 4.1 follows from Theorem 4.3 with this supplement.

Note that three rules like $AB \to A'B$, $A'B \to A'D$, and $A'D \to CD$ are not enough to replace the rules $AB \to CD$. For example, if we have the rules $S \to AB$, $B \to DE$, $AB \to CD$ in the original grammar, then this replacement makes the derivation

$$S \Rightarrow AB \Rightarrow A'B \Rightarrow A'DE \Rightarrow CDE$$

possible, though CDE is not derivable in the original grammar. Such parasitical derivations are often quite difficult to avoid when we are trying to transform a grammar into an equivalent one.

4.3 ONE-SIDED CONTEXT-SENSITIVE GRAMMARS

It can be observed that in the context-sensitive refinement of the Kuroda normal form we have only a one-sided context in each rule, but both versions, i.e., left-context and right-context, occur in some rules. The question arises whether we can generate every context-sensitive language just by, say, left-context rules. This problem had been open for a long time and a positive answer has been established in a rather complicated way by M. Pentonnen [1974a] using the ideas of Haines, Gladkij, and others. It is easier to prove the weaker result that left-context sensitive rules plus inverting rules with form $AB \to BA$ are sufficient to generate every context-sensitive language.

Theorem 4.4 For every λ-free context-sensitive grammar we can give an equivalent grammar whose rules have the forms

1) $A \to a$
2) $A \to B$
3) $A \to BC$
4) $AB \to AC$
5) $AB \to BA$

PROOF Having established the refined version of Kuroda normal form, we have to eliminate only the right-sensitive rules of form $AB \rightarrow CB$. This can be done by using the set of rules

$$
\begin{array}{ll}
AB \rightarrow AB' & A \rightarrow A' \\
A'B' \rightarrow A'D & A' \rightarrow E \\
ED \rightarrow DE & D \rightarrow C' \\
C'E \rightarrow C'E' & C' \rightarrow C \\
CE' \rightarrow CB &
\end{array}
$$

where A', B', C', D, E, and E' are new variables.

First, it is easy to see that the replacement represented by the rule $AB \rightarrow CB$ can be performed by using the new rules. Conversely, we have to show that every terminal word derivable with the aid of the new rules is also derivable in the original grammar. For this purpose let us consider all possible derivations from AB using only the new rules. (See Figure 4.1.) As can be seen in the figure, every node of this derivation-graph, except for the starting and the leftmost ending ones, is labeled by a pair of symbols containing at least one of the new variables. We shall see, further, that we cannot get rid of these new variables if we deviate from the leftmost path or do not follow it up to the end. Namely, an inconvenient (interfering) replacement of these variables may occur only in two ways:

1) Two pairs of variables shown in Figure 4.1 occur next to each other in a word derivable from S. The last variable of the first pair followed by the first variable of the second pair is equal to the left-hand side of a rule.

2) A nested pair appears in a derivation with interfering variables inside the nest. This may happen to a pair containing one of the original variables A, B, or C if the word AB is derivable from that variable in the original grammar. Such pairs are $A'B$, $A'C$, EB, EC, AB', CE, and CE'.

All these possibilities are shown in Figure 4.2, where each row of the table corresponds to a variable that occurs in the second position of a pair or in the first position of a nest, while each column corresponds to a variable that occurs in the first position of a pair or in the second position of a nest. Here the minus sign means that no interference can take place between the two variables while "yes" means that interference exists.

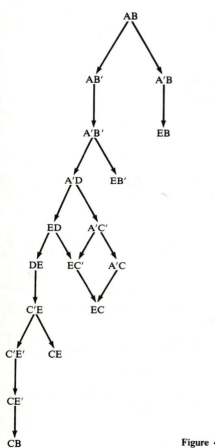

Figure 4.1

According to the five cases shown in Figure 4.2, the following subcases arise:

a) $A'C'EB$ $\quad$ $A'C'EB'$ $\quad$ $A'C'EC$ $\quad$ $A'C'ED$
$\quad$ $EC'EB$ $\quad$ $EC'EB'$ $\quad$ $EC'EC$ $\quad$ $EC'ED$
b) $C'EDE$ $\quad$ $CEDE$ $\quad$ $DEDE$
c) $A'CE'$ $\quad$ ECE'
d) $AC'E$ $\quad$ $EC'E$
e) EDE

	A	C	A'	C'	E	D	B'	E	E'
B	–	–	–	–	–	–	–	–	–
C	–	–	–	–	–	–	–	–	yes
B'	–	–	–	–	–	–	–	–	–
C'	–	–	–	–	yes	–	–	yes	–
D	–	–	–	–	–	–	–	–	–
E	–	–	–	–	–	yes	–	–	–
E'	–	–	–	–	–	–	–	–	–
A'	–	–	–	–	–	–			
E	–	–	–	–	–	yes			

Figure 4.2

After the application of the rules in question we get:

a') $A'C'E'B$, $A'C'E'B'$ $A'C'E'C$ $A'C'E'D$
    $EC'E'B$ $EC'E'B$ $EC'E'C$ $EC'E'D$

b') $C'DEE$ $CDEE$ $DDEE$

c') $A'CB$ ECB

d') $A'C'E'$ $EC'E'$

e') DEE

For case a' it is easy to see that A' can be replaced only by E, and the leading E cannot be replaced by another variable in these words. A similar situation holds for cases c' and d'. For cases b' and c', the rightmost E cannot be replaced by another variable.

The remaining variables A' and E can be interfered with by other symbols only in the same way as already discussed; this completes the proof.

One-sided context-sensitive grammars can be used as natural tools for processing words from left to right even if the language is context-free, as is the case in the following example.

Example 4.1 Let $V_N = \{S, A, B, C, D, E, H, K\}$, $V_T = \{a, +, -, \times, /, (,)\}$ and the rules be the following:

$$S \to A \qquad C) \to D) \qquad B \to (A$$

$$A \to B) \qquad CH \to DH \qquad H \to +$$

$$A \to a \qquad D \to HA \qquad H \to -$$

$$B \to (C \qquad D \to EA \qquad K \to \times$$

$$B \to BC \qquad E \to DK \qquad K \to /$$

This is a right-sensitive grammar that generates the closed arithmetic expressions where the operands are all denoted by a. We recommend that the reader try to prove this and see how this grammar reflects the priority of the infix operators (i.e., the fact that addition and subtraction are preceded by multiplication and division) overriding the simple left-to-right order. In fact, A denotes an operand or a closed term, C denotes an open term, D denotes a signed factor, and the order of the derivation represents the order of the arithmetic operations in the reverse sense.

Incidentally, the above grammar is in a normal form with four kinds of rules:

1) $A \rightarrow a$
2) $A \rightarrow B$
3) $A \rightarrow BC$
4) $AB \rightarrow CB$

Theorem 4.5 For every left-sensitive grammar, we can give an equivalent grammar in the similar normal form with the rule

4') $AB \rightarrow AC$

PROOF We may assume that terminal symbols occur only in form 1 rules. An arbitrary left-sensitive rule

$$X_1 \cdots X_m A \rightarrow X_1 \cdots X_m B_1 \cdots B_n$$

can be replaced by the set of rules as follows:

$$X_1 X_2 \rightarrow X_1 Z_2, \quad Z_2 X_3 \rightarrow Z_2 Z_3, \ldots, Z_{m-1} X_m \rightarrow Z_{m-1} Z_m$$
$$Z_m A \rightarrow Z_m Y_1, \quad Y_1 \rightarrow B_1 Y_2, \ldots, Y_{n-1} \rightarrow B_{n-1} B_n$$
$$Z_2 \rightarrow X_2, \quad Z_3 \rightarrow X_3, \ldots, Z_m \rightarrow X_m$$

where $Z_2, \ldots, Z_m$ and $Y_1, \ldots, Y_{n-1}$ are newly introduced variables.

A similar normal form can be obtained, of course, for right-sensitive grammars. As we have mentioned without proof, left-sensitive grammars are of the same generating power as arbitrary context-sensitive ones, so the above normal form is universal for type 1 languages. It is, however, extremely difficult to obtain this normal form for an arbitrary context-sensitive grammar and thus it bears merely theoretical significance. It is interesting to note that we can give very simple one-sided context-sensitive grammars that generate non-context-free languages.

Example 4.2 Let $V_N = \{S, A_1, A_2, B_1, B_2, C_1, C_2\}$, $V_T = \{a, b, c\}$, and let us have these rules:

$S \rightarrow A_1 B_1 C_1$

$A_1 \rightarrow aA_2 B_2$	$B_2 B_1 \rightarrow B_2 B_2$	$B_2 C_1 \rightarrow B_2 C_2 c$
$A_2 \rightarrow aA_1 B_1$	$B_1 B_2 \rightarrow B_1 B_1$	$B_1 C_2 \rightarrow B_1 C_1 c$
$A_1 \rightarrow a$	$B_1 \rightarrow b$	$C_1 \rightarrow c$
$A_2 \rightarrow a$	$B_2 \rightarrow b$	$C_2 \rightarrow c$

It is not difficult to see that this grammar generates the language $\{a^n b^n c^k \mid n \geq k \geq 1\}$. This follows from the fact that every application of the rule $A_1 \to aA_2B_2$ or $A_2 \to aA_1B_1$ sends a signal through the strings of B's to the C_1 or C_2 on their right. This signal may be killed on its way or it may reach the C_i where it makes possible to deposit a small c.

EXERCISES

4.1 Convert the grammar given in Exercise 1.5 into the refined version of Kuroda normal form.

4.2 Convert the grammar of Example 4.2 into the normal form established by Theorem 4.5.

4.3 Find a context-free grammar which is equivalent to the grammar given in Example 4.1.

4.4 Find a context-sensitive grammar which is equivalent to the grammar given in Example 1.2.

4.5 Find a context-sensitive grammar to generate the language $\{a^m b^n c^m d^n \mid m, n \geq 1\}$.

UNRESTRICTED PHRASE STRUCTURE LANGUAGES

5.1 A NORMAL FORM FOR TYPE 0 GRAMMARS

A generalization of Kuroda normal form can be developed for type 0 grammars. This normal form will be very useful in Chapter 6, where we shall establish the relationship between grammars and automata.

Theorem 5.1 For every type 0 grammar there is an equivalent grammar G' in which every rule has any of the following forms:

1) $S \to \lambda$
2) $A \to a$
3) $A \to B$
4) $A \to BC$
5) $AB \to AC$
6) $AB \to CB$
7) $AB \to B$

where the initial symbol S may occur only on the left-hand sides of the rules.

PROOF First, we replace every rule of form $P \to \lambda$ by the set of rules $PX \to X$ and $XP \to X$ for all X in $V_N \cup V_T$. This way we get a grammar G' for which $L(G') = L(G) - \{\lambda\}$. Now, if $\lambda \in L(G)$, then we add the rule $S' \to \lambda$. Next, we transform G' such that terminal symbols will occur in form 2 rules only. All the other rules will then have form $P \rightarrow Q$ with $P, Q \in V_N^+$. For a length-increasing rule of this form with $|P| \leqslant |Q|$ we apply the same method as we have used for the Kuroda normal form. Thus, we have to consider only length-decreasing rules here.

A length-decreasing rule has the form

$$X_1 \cdots X_m \rightarrow Y_1 \cdots Y_n \qquad (m > n \geqslant 1)$$

where X_i and Y_j are in V_N. Such a rule will be replaced by the set of rules:

$$X_{m-1}X_m \rightarrow Z_m U_m \qquad\qquad Z_m U_m \rightarrow U_m$$
$$X_{m-2}U_m \rightarrow Z_{m-1}U_{m-1} \qquad Z_{m-1}U_{m-1} \rightarrow U_{m-1}$$
$$\vdots \qquad\qquad\qquad\qquad\qquad \vdots$$
$$X_n U_{n+2} \rightarrow Z_{n+1}U_{n+1} \qquad Z_{n+1}U_{n+1} \rightarrow U_n Y_n$$
$$X_{n-1}U_n \rightarrow U_{n-1}Y_{n-1}$$
$$\vdots$$
$$X_1 U_2 \rightarrow U_1 Y_1$$
$$U_1 Y_1 \rightarrow Y_1$$

where $U_1, \ldots, U_m$ and $Z_{n+1}, \ldots, Z_m$ are newly introduced variables. Here we have $AB \rightarrow CD$ type rules each of which can be replaced by four rules of form 5 and 6, respectively. This way we can obtain a grammar in the above normal form which generates the same language; this completes the proof.

REMARK In contrast to our previous proofs the proof of Theorem 5.1 contains a nonconstructive step which is based on a merely existential statement. Namely, we do not know how to decide in general whether or not λ is in $L(G)$. If the grammar has no $P \rightarrow \lambda$ type rules, then obviously $\lambda \notin L(G)$. But, if it does have λ-rules then λ may or may not belong to $L(G)$. So we can always construct a grammar G' in normal form such that $L(G') = L(G) - \{\lambda\}$, but we do not know whether we should include the rule $S' \rightarrow \lambda$ or not. This is why Theorem 5.1 has been formulated as "there is an equivalent grammar" without saying anything about the possibility of constructing such a grammar. Existence theorems like that are based on the logical principle of the "excluded middle", which means that either $\lambda \in L(G)$ or $\lambda \notin L(G)$ must be true, and no other possibility exists, even if we cannot decide which of the two cases occurs. As we shall see later, it is indeed undecidable whether λ is in $L(G)$ for an arbitrary type 0 grammar G. Of course, this would not mean that $\lambda \in L(G)$ is undecidable in every case.

For type 1 grammars, we have seen that $P \in L(G)$ is always decidable. (See Theorem 4.2.) For type 0 grammars we cannot apply the same reasoning, since we also have length-increasing and length-decreasing rules, so we cannot easily establish an upper bound to the number of steps in a derivation. The number of possible steps can be reduced by considering only leftmost derivations. For this purpose we have to generalize the notion of a leftmost derivation.

Definition 5.1 A derivation in a type 0 grammar

$$W_0 \Rightarrow W_1 \Rightarrow \cdots \Rightarrow W_n$$

is called *leftmost* if for $i = 0, 1, \ldots, n - 1$ there are words P_i, Q_i, X_i, and Y_i in $(V_N \cup V_T)^*$ such that $P_i \to Q_i \in F$ and

$$W_i = X_i P_i Y_i \qquad W_{i+1} = X_i Q_i Y_i$$

where $|X_i| < |X_{i+1} P_{i+1}|$.

In other words, a derivation is leftmost if in each step the rewriting occurs to the right of, or overlapping with, that part of the word which was rewritten just before.

Theorem 5.2 For every derivation $P \overset{*}{\Rightarrow} Q$ in a type 0 grammar G we can give a leftmost derivation $P \overset{*}{\underset{\text{left}}{\Rightarrow}} Q$ with the same number of steps.

PROOF Suppose that the derivation

$$P = W_0 \Rightarrow W_1 \Rightarrow \cdots \Rightarrow W_n = Q$$

is not leftmost and let k be here the largest index such that

$$W_0 \overset{*}{\Rightarrow} W_k$$

is leftmost. Then we have

$$W_k = X_k P_k Y_k$$
$$W_{k+1} = X_k Q_k Y_k = X_{k+1} P_{k+1} Y_{k+1}$$

and

$$W_{k+2} = X_{k+1} Q_{k+1} Y_{k+1}$$

such that $P_k \to Q_k \in F$, $P_{k+1} \to Q_{k+1} \in F$ and

$$|X_k| \geqslant |X_{k+1} P_{k+1}|$$

But then $X_k = X_{k+1} P_{k+1} R_k$ for some $R_k \in (V_N \cup V_T)^*$, and thus the order of the k-th and the $k + 1$-st steps can be inverted which makes the derivation $W_0 \overset{*}{\Rightarrow} W_{k+1}$ leftmost. Continuing this process we can make the entire derivation leftmost, which completes the proof.

Note that the leftmost derivation in the above sense is also unique as can be verified easily by the reader.

Example 5.1 The language $\{PP | P \in \{a, b\}^*\}$ can be generated by the grammar $G = (\{S, A, B, C\}, \{a, b\}, S, F)$ where F contains the following rules:

$$S \to ASa \qquad AC \to aC \qquad BC \to bC$$
$$S \to BSb \qquad Aa \to aA \qquad Ba \to aB$$
$$S \to C \qquad Ab \to bA \qquad Bb \to bB$$
$$C \to \lambda$$

In this grammar we can first derive a string of the form $Q^{-1}SP$ where $P \in \{a, b\}^*$ and Q is just the capitalized version of P. Then S can be changed into C, which will act as a reflecting and "decapitalizing" mirror for Q^{-1}. Specifically, an A or B adjacent to C can be changed into the corresponding small letter. Further, all A's and B's can travel to the right across terminal symbols until they reach C. When they hit C, they bounce back and drop dead. (In other words, they turn into the corresponding terminal symbol.) This way their order will be reversed, giving PCP and finally PP. (If $C \to \lambda$ is applied too soon, then the remaining A's and B's cannot be eliminated.)

Now, in order to convert this grammar into the normal form of Theorem 5.1, we replace $C \to \lambda$ by three rules:

$$S \to \lambda \qquad Ca \to a \qquad Cb \to b$$

No other rules are needed here, because C must be followed by an a or b, except for the derivation of λ. Next we introduce four new variables D, E, X, and Y, to obtain the following rules:

$$S \to \lambda \qquad AC \to XC \qquad BC \to YC$$
$$S \to AD \qquad AX \to XA \qquad BX \to XB$$
$$S \to BE \qquad AY \to YA \qquad BY \to YB$$
$$D \to SX \qquad CX \to X \qquad CY \to Y$$
$$E \to SY \qquad X \to a \qquad Y \to b$$

The last step in getting the normal form is replacing each inverting rule like $AX \to XA$ by four context-sensitive rules, as in the refined version of Kuroda normal form. (Note that in this example we did not have to transform any length-decreasing rules.)

5.2 DERIVATION GRAPH

In Section 3.2 we have seen that context-free derivations can be represented by directed trees. This representation cannot be used for derivations in type 1 or type 0 grammars because here we have more than one symbol on the left-hand side of the rules. It is, however, easy to overcome this difficulty in the following way:

Given a phrase-structure grammar $G = (V_N, V_T, S, F)$ we assign a unique name to each production rule in F; in symbols

$$f: P \rightarrow Q \in F$$

where P and Q are in $(V_N \cup V_T)^*$ while $f \notin (V_N \cup V_T)$ is the name of the rule. Hence, the set of rules can be represented by the set of names

$$V_F = \{f_1, f_2, \ldots, f_n\}$$

where each f_i stands for a specific production rule assigned to it. It is assumed that $V_F \cap (V_N \cup V_T) = \varnothing$. The graph of a derivation will then be constructed

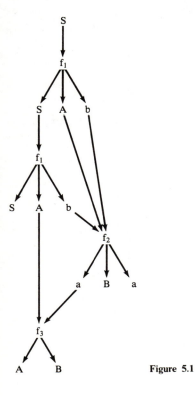

Figure 5.1

by introducing a new node labeled by the production name for each application of a rule. This node will be entered by edges coming from the nodes labeled by the terminal and nonterminal symbols occurring on the left-hand side of the rule while outgoing edges will be drawn to the nodes with labels corresponding to the right-hand side of the rule.

Let us have, for instance, a grammar $G = (V_N, V_T, S, F)$, where $V_N = \{S, A, B\}$, $V_T = \{a, b\}$ and let $V_F = \{f_1, f_2, f_3, f_4\}$ while the productions in question are as follows:

$$f_1: S \rightarrow SAb$$

$$f_2: bAb \rightarrow aBa$$

$$f_3: Aa \rightarrow AB$$

$$f_4: BB \rightarrow aBb$$

Then the graph of Figure 5.1 on the previous page represents the derivation

$$S \Rightarrow SAb \Rightarrow SAbAb \Rightarrow SAaBa \Rightarrow SABBa$$

Every derivation is represented this way by a unique graph, and conversely, every derivation graph represents a unique leftmost derivation. The last statement is not quite obvious, but can be proved without difficulty.

EXERCISES

5.1 Find a context-sensitive grammar to generate the language given in Example 5.1 *Hint:* The length-decreasing rules, $CX \rightarrow X$ and $CY \rightarrow Y$, can be eliminated if the first symbol to the right of the center is used as the mirror.

5.2 Find a derivation graph for the string *abbaabba* in each of the equivalent grammars generating the language given in Example 5.1.

CHAPTER SIX

AUTOMATA AND THEIR LANGUAGES

So far, we have used phrase structure grammars for generating formal languages. It is, however, quite reasonable to define a formal language with the aid of some mechanical device which is capable of processing strings of symbols. Having such a device, called an *automaton*, with two possible reactions ("yes" or "no") for any input string presented to it, we say that it *accepts* or *recognizes* the language consisting of all words provoking the answer "yes." The words getting the answer "no" are said to be *rejected* by the automaton.

We hasten to mention that it is quite realistic to consider the possibility of getting no answer at all. In computer programming we are often faced with the problem of running a program in an infinite loop.

The behavior of automata depends on their structure, which we shall study next. We shall focus our attention on language-accepting problems. Automata will thus be considered mainly as language-accepting devices and their internal structure will be discussed only to the extent of its relation to the accepted language. In our opinion, generative grammars, on one hand, and automata, on the other hand, correspond to the two aspects of the usage of a language. Namely, the speaker (source) would normally use some generative grammar as a synthetical device while the listener (receiver) needs a recognizer, i.e., some analytical device. Of course, as we shall see later, generative grammars can also be used for analyzing sentences and automata can also be designed to generate sentences of a language, but we consider the former approach as the basic one.

In this chapter we will show that there is a fairly natural hierarchy of automata that corresponds exactly to the Chomsky hierarchy of languages.

6.1 FINITE AUTOMATA

The theory of finite automata is a highly developed mathematical theory. As we have said before, we shall discuss here only those features which are necessary for the theory of formal languages.

Definition 6.1 A finite automaton is an ordered quintuple

$$A = (K, T, M, q_0, H)$$

where K is a finite nonempty set of *states*

T is a finite alphabet of *input symbols*

M is a mapping called the *transition function* from set $K \times T$ to K

$q_0 \in K$ is the *initial state*

$H \subseteq K$ is a set of *accepting states*

The way of working of a finite automaton is to be understood as making brisk moves at discrete time intervals. Each move consists of reading the next input symbol and entering the new state selected by the transition function applied to the argument pair of the current state and the current input symbol. We may assume that the input symbols are presented one at a time to the automaton via some input tape that is divided into squares, each holding a single symbol. In each move the input tape is advanced by one square in front of a read head connected to the finite state control device (see Figure 6.1).

Initially the finite automaton is in state q_0 and the read head is scanning the first symbol of the input word $P \in T^*$ written on the input tape. Then it makes a sequence of moves while reading the input tape. If it enters a state in H just after reading the last symbol of P, then P is accepted by the automaton; otherwise (i.e., in case of entering eventually a state in $K - H$) P is rejected.

As can be seen from this definition finite automata may never get into infinite loops as long as the input is finite, since they have to read a new

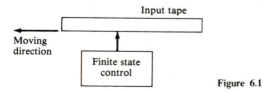

Figure 6.1

symbol for each move. Therefore, they must produce a definite answer in $|P|$ moves.

Example 6.1 Let $A = (K, T, M, q_0, H)$ be a finite automaton with $K = \{q_0, q_1, q_2, q_3\}$, $T = \{a, b\}$, $H = \{q_0\}$, and let the mapping M be defined as follows:

$$M(q_0, a) = q_2 \qquad M(q_0, b) = q_1$$
$$M(q_1, a) = q_3 \qquad M(q_1, b) = q_0$$
$$M(q_2, a) = q_0 \qquad M(q_2, b) = q_3$$
$$M(q_3, a) = q_1 \qquad M(q_3, b) = q_2$$

It is more convenient to specify the mapping M with the aid of the transition table given in Table 6.1, where each row corresponds to a state and each column represents an input symbol. The entries in the table are the values of M.

Let us start this finite automaton from state q_0 with the input word *bbabab*. Clearly, this word will be accepted by the automaton, because it will enter the states $q_1, q_0, q_2, q_3, q_1, q_0$ in that order and the state q_0 is an accepting state in this example. It can be shown easily that this automaton accepts exactly those words in $\{a, b\}^*$ which have an even number of a's and also an even number of b's. This fact may be concluded more easily from the so-called transition graph, where each state is represented by a node and each directed edge represents a transition from one state to the other in case of reading the input symbol labeling the edge (see Figure 6.2).

Transition graphs and transition tables can be used equivalently to define the mapping M. The initial state and the accepting states may be distinguished by additional marks if so desired.

Here we have assumed that the mapping M is a one-valued function, which means that we have a unique new state for each pair (q, a) in $K \times T$. In this case the automaton is called *deterministic*, because the sequence of moves is completely determined by the input word.

If, however, we allow for multivalued transition functions as well, then we get the notion of the *nondeterministic* finite automaton, where $M(q, a)$

Table 6.1

M	a	b
q_0	q_2	q_1
q_1	q_3	q_0
q_2	q_0	q_3
q_3	q_1	q_2

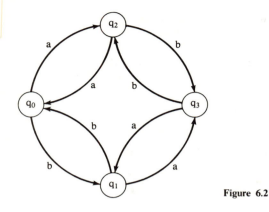

Figure 6.2

is a subset of K rather than a unique member of it. Therefore, the current state of a nondeterministic finite automaton is to be considered as any of the states in a subset of K, rather than a unique state. Thus the initial state is also replaced by a set of initial states $K_0 \subseteq K$. It may also happen that for an input symbol a the set $M(q, a)$ is empty for all q in the set of current states. In that case the automaton gets stuck.

The mapping M can also be given in terms of rewriting rules such that

$$qa \to p \in M \qquad \text{iff} \qquad p \in M(q, a)$$

If M contains exactly one rule with the left-hand side qa for each pair (q, a) in $K \times T$, then the finite automaton is deterministic; otherwise it is nondeterministic.

Definition 6.2 Given a finite automaton $A = (K, T, M, K_0, H)$ and two words X, Y in KT^* (where KT^* denotes the catenation of the sets K and T^*), then we say that the automaton A *reduces* X to Y in one move, (in symbols $X \underset{A}{\Rightarrow} Y$) iff there is a move $qa \to p$ in M and a word $P \in T^*$ such that $X = qaP$ and $Y = pP$.

Definition 6.3 A finite automaton A reduces a word $X \in KT^*$ to a word $Y \in KT^*$ (in symbols $X \underset{A}{\overset{*}{\Rightarrow}} Y$) iff $X = Y$ or else there is some $Z \in KT^*$ such that $X \underset{A}{\overset{*}{\Rightarrow}} Z$ and $Z \underset{A}{\Rightarrow} Y$.

REMARK The relation $\underset{A}{\Rightarrow}$ and its transitive and reflexive closure $\underset{A}{\overset{*}{\Rightarrow}}$ are essentially the same as the derivation in a grammar. We want to emphasize this similarity by using similar notations.

Definition 6.4 The language accepted by a finite automaton A is defined as

$$L(A) = \left\{ P \in T^* | q_0 P \overset{*}{\underset{A}{\Rightarrow}} p \quad \text{for some } q_0 \in K_0 \text{ and } p \in H \right\}$$

Note that the empty word λ is in $L(A)$ if and only if $K_0 \cap H \neq \varnothing$. Each move $qa \rightarrow p$ in M is a length-decreasing rewriting rule but it is not difficult to realize that reductions correspond to reverse derivations in a left-linear grammar.

Theorem 6.1 For every nondeterministic finite automaton A, we can give a type 3 grammar G such that $L(G) = L(A)$.

PROOF Let $A = (K, T, M, K_0, H)$ be a finite automaton. We define the grammar $G = (V_N, V_T, S, F)$ such that $V_N = K \cup \{S\}$, $V_T = T$, and

1) $p \rightarrow a \in F$ iff $q_0 a \rightarrow p \in M$ for some $q_0 \in K_0$
2) $p \rightarrow qa \in F$ iff $qa \rightarrow p \in M$
3) $S \rightarrow p \in F$ iff $p \in H$
4) $S \rightarrow \lambda \in F$ iff $K_0 \cap H \neq \varnothing$

It is obvious that $\lambda \in L(G)$ iff $\lambda \in L(A)$. Let $P \in L(A)$ and $P \neq \lambda$. Then we have some $q_0 \in K_0$ and $p \in H$ such that $q_0 P \overset{*}{\underset{A}{\Rightarrow}} p$. Following this reduction in the reverse sense, we can construct the derivation $p \underset{G}{\Rightarrow} q_0 P$ using rules from group 2. The last step in that derivation is the application of a rule with form $p_1 \rightarrow q_0 a$ where $q_0 \in K_0$. Therefore, we have a related rule $p_1 \rightarrow a$ in group 1, which gives $p \overset{*}{\underset{G}{\Rightarrow}} P$. Finally, we take $S \rightarrow p$ from group 3 and thus,

$$S \underset{G}{\Rightarrow} p \overset{*}{\underset{G}{\Rightarrow}} P$$

is established, which proves that $L(A) \subseteq L(G)$.

Conversely, let $P \in L(G)$ and $P \neq \lambda$, that is, $S \overset{*}{\underset{G}{\Rightarrow}} P$ and $P \in T^+$. By the construction of G this derivation must be of the form

$$S \underset{G}{\Rightarrow} p \overset{*}{\underset{G}{\Rightarrow}} p_1 Q \underset{G}{\Rightarrow} aQ = P$$

with $p \in H$, and $p_1 \rightarrow a \in F$, which implies that

$$q_0 P \overset{*}{\underset{A}{\Rightarrow}} p$$

and, thus, $P \in L(A)$.

The rules in G are left-linear, but we know by Theorem 3.7 that we can find an equivalent right-linear grammar, which completes the proof.

Another way of proving Theorem 6.1 would be to construct a right-linear grammar that simulates the finite automaton. This idea is used in the proof of the converse theorem.

Theorem 6.2 For every type 3 grammar G we can give a finite automaton A such that $L(A) = L(G)$.

PROOF Without the loss of generality we can assume that $G = (V_N, V_T, S, F)$ is in the normal form established by Theorem 3.8. We construct $A = (K, T, M, q_0, H)$ such that $K = V_N$, $T = V_T$, $q_0 = S$, and

$$H = \{Z \in V_N | Z \to \lambda \in F\}$$

while the mapping M is defined as follows

$$Xa \to Y \in M \qquad \text{iff} \qquad X \to aY \in F$$

It is easy to see that to every derivation $S \overset{*}{\underset{G}{\Rightarrow}} P$ in G we have a sequence of moves in A of the form

$$SP \overset{*}{\underset{A}{\Rightarrow}} Z$$

with $Z \in H$. Also, conversely, to every reduction of the latter form in A we can find a corresponding derivation in G, which completes the proof.

Example 6.2 Consider the right-linear grammar

$$G = (\{S, Z\}, V_T, S, F)$$

where $V_T = \{a, b, c, \ldots, z, 0, 1, 2, \ldots, 9\}$ and the set of rules in F are

$$S \to aZ, S \to bZ, \ldots, S \to zZ$$
$$Z \to aZ, Z \to bZ, \ldots, Z \to zZ$$
$$Z \to 0Z, Z \to 1Z, \ldots, Z \to 9Z$$
$$Z \to \lambda$$

This grammar generates the strings which start with a letter followed by zero or more letters or digits. (Such strings are used as identifiers in most programming languages.) The finite automaton $A = (\{S, Z\}, V_T, M, S, \{Z\})$ with the mapping M given in Table 6.2 will accept precisely the same language.

This finite automaton is deterministic, although incomplete, because the mapping M is undefined for some pairs in $K \times T$. But we can easily complete the mapping M by introducing a dummy state to replace the

Table 6.2

M	a	b	c	$\cdots$	z	0	1	2	$\cdots$	9
S	Z	Z	Z	$\cdots$	Z	$\varnothing$	$\varnothing$	$\varnothing$	$\cdots$	$\varnothing$
Z	Z	Z	Z	$\cdots$	Z	Z	Z	Z	$\cdots$	Z

empty (undefined) set in the table. The dummy state U will then be a killing state from which every transition goes back to the same state. Thus, an equivalent deterministic and complete finite automaton A' can be given as

$$A' = (\{S, Z, U\}, V_T, M', S, \{Z\})$$

where the mapping M' is shown in Table 6.3.

Note that the finite automaton constructed from a right-linear grammar will in general be nondeterministic. That is, if both $A \to aB$ and $A \to aC$ belong to F for some A, a, B, and C with $B \neq C$ then the construction gives us $M(A, a) = \{B, C\}$, i.e., $Aa \to B \in M$ and $Aa \to C \in M$ at the same time.

Thus, we can say that the class $\mathfrak{L}_3$ coincides with the class of languages accepted by nondeterministic finite automata. Nondeterministic finite automata are, in fact, very similar to type 3 grammars. Now, the question arises whether nondeterminism was essential here to establish the correspondence between finite automata and type 3 grammars. In other words, is there any type 3 language that is accepted by a nondeterministic finite automaton but cannot be accepted by any deterministic one? The answer is, fortunately, "no," which may be surprising at first, but a closer look at the nondeterminism in the finite case will make it quite natural. Namely, the set of current states in a nondeterministic finite automaton is a subset of the finite set of states. The way of working of a nondeterministic finite automaton can be imagined as the possibly scattered cloud (subset) of current states changing from time to time over the set of states K. But each subset of K can be considered as a unique state of a finite automaton having as many distinct states as is the number of the subsets of K. So we can simulate the behavior of the nondeterministic

Table 6.3

M'	a	b	c	$\cdots$	z	0	1	2	$\cdots$	9
S	Z	Z	Z	$\cdots$	Z	U	U	U	$\cdots$	U
Z	Z	Z	Z	$\cdots$	Z	Z	Z	Z	$\cdots$	Z
U	U	U	U	$\cdots$	U	U	U	U	$\cdots$	U

automaton by a deterministic one which has many more states. Nondeterminism can thus be considered here as a merely technical device for making the proofs easier. For the sake of completeness, we shall present a formal proof of the equivalence, but first let us see an example.

Example 6.3 Let $A = (K, T, M, K_0, H)$ be a nondeterministic finite automaton with

$$K = \{0, 1, 2\} \qquad T = \{a, b\} \qquad K_0 = \{0\} \qquad H = \{2\}$$

and let M be defined as follows:

$$M(0, a) = \{0, 1\} \qquad M(0, b) = \{1\}$$
$$M(1, a) = \varnothing \qquad M(1, b) = \{2\}$$
$$M(2, a) = \{0, 1, 2\} \quad M(2, b) = \{1\}$$

The corresponding deterministic finite automaton will have the states

$$\varnothing, \{0\}, \{1\}, \{2\}, \{0, 1\}, \{0, 2\}, \{1, 2\}, \{0, 1, 2\}$$

and the mapping M' will be defined as

$$M'(\varnothing, a) \quad = \varnothing \qquad M'(\varnothing, b) \quad = \varnothing$$
$$M'(\{0\}, a) \quad = \{0, 1\} \qquad M'(\{0\}, b) \quad = \{1\}$$
$$M'(\{1\}, a) \quad = \varnothing \qquad M'(\{1\}, b) \quad = \{2\}$$
$$M'(\{2\}, a) \quad = \{0, 1, 2\} \qquad M'(\{2\}, b) \quad = \{1\}$$
$$M'(\{0, 1\}, a) \quad = \{0, 1\} \qquad M'(\{0, 1\}, b) \quad = \{1, 2\}$$
$$M'(\{0, 2\}, a) \quad = \{0, 1, 2\} \qquad M'(\{0, 2\}, b) \quad = \{1\}$$
$$M'(\{1, 2\}, a) \quad = \{0, 1, 2\} \qquad M'(\{1, 2\}, b) \quad = \{1, 2\}$$
$$M'(\{0, 1, 2\}, a) = \{0, 1, 2\} \qquad M'(\{0, 1, 2\}, b) = \{1, 2\}$$

while the set of accepting states will be

$$H' = \{\{2\}, \{0, 2\}, \{1, 2\}, \{0, 1, 2\}\}$$

since we have $H = \{2\}$. (See the construction in Theorem 6.3.)

The states in K' can be numbered by integers from 0 to 7 or by any eight different symbols, and this will not influence the accepted language. Note that M' is defined also for the empty set $\varnothing$ so the deterministic automaton A' never gets stuck although the state $\varnothing$ represents here a killing state which, if entered once, cannot be left any more no matter what input follows.

Theorem 6.3 For every nondeterministic finite automaton $A = (K, T, M, K_0, H)$ we can give a deterministic finite automaton $A' = (K', T, M', q'_0, H')$ such that $L(A) = L(A')$.

Proof Let K' be the set of all subsets of K. (If K has k elements, then K' has 2^k members.) We define the mapping M' for all $q' \in K'$ and $a \in T$ as follows:

$$M'(q', a) = \bigcup_{q \in q'} M(q, a)$$

In detail $M'(\langle q_i, \ldots, q_j \rangle, a) = M(q_i, a) \cup \cdots \cup M(q_j, a)$.
Further, let $q'_0 = K_0$ and

$$H' = \{q' \in K' | q' \cap H \neq \varnothing\}.$$

In order to show that $L(A) \subseteq L(A')$ we prove a lemma first.

Lemma A For every $p, q \in K$, $q' \in K'$, and $P, Q \in T^*$, if

$$qP \underset{A}{\overset{*}{\Rightarrow}} pQ \quad \text{and} \quad q \in q'$$

then there is a p' in K' such that

$$q'P \underset{A'}{\overset{*}{\Rightarrow}} p'Q \quad \text{and} \quad p \in p'$$

This will be shown by induction on the number of moves involved in the reduction $qP \underset{A}{\overset{*}{\Rightarrow}} pQ$. For zero moves the assertion is trivial. Assume that it is true for n moves ($n \geq 0$) and let $qP \underset{A}{\overset{*}{\Rightarrow}} pQ$ have $n + 1$ moves. Then we have some $q_1 \in K$ and $P_1 \in T^*$ such that

$$qP \underset{A}{\Rightarrow} q_1 P_1 \underset{A}{\overset{*}{\Rightarrow}} pQ$$

Hence, there is an $a \in T$ with $P = aP_1$ and $q_1 \in M(q, a)$. But $M(q, a) \subseteq M'(q', a)$ for $q \in q'$, and thus we can choose $q'_1 = M'(q', a)$ which yields

$$q'P \underset{A'}{\Rightarrow} q'_1 P_1 \quad \text{with} \quad q_1 \in q'_1$$

Now the induction hypothesis gives us some $p' \in K'$ with

$$q'_1 P_1 \underset{A'}{\overset{*}{\Rightarrow}} p'Q \quad \text{and} \quad p \in p'$$

which completes the proof of Lemma A.

Assume that $P \in L(A)$, that is, $q_0 P \underset{A}{\overset{*}{\Rightarrow}} p$ for some $q_0 \in K_0$ and $p \in H$. Then by Lemma A we have some p' such that $q'_0 P \underset{A'}{\overset{*}{\Rightarrow}} p'$ and $p \in p'$. But by the definition of H', $p \in p'$ and $p \in H$ implies that $p' \in H'$, which gives the result.

To show that $L(A') \subseteq L(A)$ we prove the converse of Lemma A.

Lemma B For every $p', q' \in K'$, $p \in K$ and $P, Q \in T^*$ if

$$q'P \overset{*}{\underset{A'}{\Rightarrow}} p'Q \quad \text{and} \quad p \in p'$$

then there is a q in K such that

$$qP \overset{*}{\underset{A}{\Rightarrow}} pQ \quad \text{and} \quad q \in q'$$

Again we use induction on the number of moves. For zero moves the assertion is trivial. Otherwise $q'P \overset{*}{\underset{A'}{\Rightarrow}} p'Q$ has form

$$q'P \overset{*}{\underset{A'}{\Rightarrow}} p_1'Q_1 \underset{A'}{\Rightarrow} p'Q$$

where $Q_1 = aQ$ for some $a \in T$ and $p_1' \in K'$. Then clearly

$$p \in p' = M'(p_1', a) = \bigcup_{p_1 \in p_1'} M(p_1, a)$$

so we must have some p_1 in p_1' for which $p \in M(p_1, a)$ holds. For such a p_1 we have

$$p_1 Q_1 \underset{A}{\Rightarrow} pQ$$

and by the induction hypothesis this gives some q in q' with

$$qP \overset{*}{\underset{A}{\Rightarrow}} p_1 Q_1$$

which completes the proof of Lemma B.

Now let $q_0'P \overset{*}{\underset{A'}{\Rightarrow}} p'$ and $p' \in H'$. Then by definition of H' we have some $p \in p'$ with $p \in H$, and thus by Lemma B we get some $q_0 \in q_0'$ such that $q_0 P \overset{*}{\underset{A}{\Rightarrow}} p$. This means that $L(A') \subseteq L(A)$, and so the proof is completed.

With the aid of deterministic finite automata we can prove a number of important properties of type 3 languages.

Theorem 6.4 The class of languages $\mathcal{L}_3$ is closed under complementation.

PROOF We have seen that every $L \in \mathcal{L}_3$ is accepted by some deterministic finite automaton. If L is accepted by the deterministic finite automaton

$$A_1 = (K, T, M, q_0, H)$$

then its complement $\bar{L} = T^* - L$ is accepted by

$$A_2 = (K, T, M, q_0, K - H)$$

which completes the proof.

Corollary A The class $\mathcal{L}_3$ forms a boolean algebra.

PROOF We have seen in Chapter 2 that $\mathcal{L}_3$ is closed under the set-theoretical union. Using the well-known set-theoretical identity

$$L_1 \cap L_2 = \overline{\bar{L}_1 \cup \bar{L}_2}$$

the closure under intersection follows from Theorem 6.4.

Corollary B For any regular language R and context-free language L it is decidable whether or not $L \subseteq R$.

PROOF Clearly, $L \subseteq R$ iff $L \cap \bar{R} = \varnothing$. But $\bar{R}$ is also type 3, hence, by Theorem 3.9, $L \cap \bar{R}$ is context-free whose emptiness problem is decidable according to Theorem 3.4.

REMARK This corollary may be of interest for compiler construction where the lexical analyzer uses a finite automaton whose language must include the original context-free language as a subset.

Corollary C It is decidable whether two arbitrary type 3 grammars generate the same language.

PROOF Let G_1 and G_2 be two type 3 grammars and $L_1 = L(G_1)$, $L_2 = L(G_2)$. Then the language

$$L_3 = (L_1 \cap \bar{L}_2) \cup (\bar{L}_1 \cap L_2)$$

is also type 3, so we can find a type 3 grammar G_3 that generates L_3. But $L_1 = L_2$ holds iff L_3 is empty, which is decidable for every type 2 language.

EXERCISES

6.1 Find a type 3 grammar that generates the language accepted by the automaton given in Example 6.1.

6.2 Develop a deterministic finite automaton that accepts the language given in Example 2.1.

6.3 Develop a deterministic finite automaton that accepts the complement of the language accepted by the automaton given in Example 6.3.

6.4 Prove that a nonempty language accepted by a deterministic finite automaton must contain some word P which is not longer than the number of the states of the automaton. *Hint*: If the input word is longer than the number of the states in K, then there must be a state which is entered twice in the course of processing that word.

6.5 Prove that the language $\{a^n b^n | n \geq 1\}$ is not type 3. *Hint*: Observe again that a deterministic finite automaton having only k different states must repeat some state when processing a word longer than k.

6.6 Find a nondeterministic finite automaton to accept the language generated by the following grammar:

$$G = (\{S, A, B, C\}, \{a, b\}, S, F)$$

where F contains the rules

$$S \rightarrow aS \qquad S \rightarrow bS \qquad S \rightarrow aA$$
$$A \rightarrow bB \qquad B \rightarrow aC \qquad C \rightarrow \lambda$$

Also, find an equivalent deterministic finite automaton.

6.7 Find a nondeterministic and then a deterministic finite automaton to accept the language of Exercise 3.3.

6.8 Find a deterministic finite automaton that accepts the set of words (finite language) as follows:

$$\{one, two, three, four, five\}$$

6.9 Find a nondeterministic finite automaton that accepts the language denoted by the regular expression

$$(a; b^*|a^*; c); b|(a|b); (c; b; c^*; a)^*$$

Hint: First find an equivalent right-linear grammar by using Theorem 2.2.

6.2 PUSHDOWN AUTOMATA

Finite automata have been shown to be equivalent to type 3 grammars. It can be guessed that the recognition of the type 2 language $\{a^n b^n\}$ requires an infinite number of states to memorize the number of a's that have been read before the first b is encountered. This means that we have to accept the idea of dealing with infinite automata. A straightforward generalization of the notion of finite automata by allowing for an infinite number of states seems to be far too general. Therefore, we shall consider specific structures for handling theoretically unlimited amounts of data in some relatively simple way.

A very important type of infinite automata is the pushdown automaton where we have a potentially infinite pushdown store in addition to the finite state control device (see Figure 6.3). The pushdown store is somewhat similar to a *stack* or pile where new data are added always on the top of previous recordings (the latter being pushed down the store) but reading occurs in the reverse order and clears the information that has been read. The contents of the pushdown store is of potentially infinite length, but only the most recently stored data can be accessed. In other words, we cannot read previously recorded information directly from the store unless we first delete all other information that has been stored later than the requested one. This storage technique is also called *first in last out* (FILO), which means that the sooner the

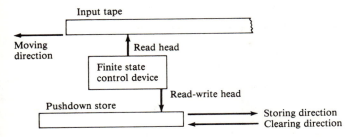

Figure 6.3 Pushdown automaton.

information is stored the later it can be retrieved. The mathematical definition of this model is given below.

Definition 6.5 A pushdown automaton is an ordered seventuple

$$A = (Z, K, T, M, z_0, q_0, H)$$

where Z is a finite alphabet of *pushdown symbols*
    K is a finite set of *internal states*
    T is a finite alphabet of *input symbols*
    M is a mapping from $Z \times K \times (T \cup \langle\lambda\rangle)$ into the finite subsets of $Z^* \times K$; the *transition function*
$z_0 \in Z$ is the *initial symbol*
$q_0 \in K$ is the *initial state*
$H \subseteq K$ is the set of *accepting states*

This definition has already been given for the nondeterministic case. The next move depends on the topmost symbol currently in the pushdown store, on the current state of the control device, and possibly on the next input symbol. Therefore, the complete state of the pushdown automaton is represented by its configuration.

Definition 6.6 The configuration of a pushdown automaton is a word of form Wq where $W \in Z^*$ is the current contents of the pushdown store and $q \in K$ is the current state of the finite control.

A nondeterministic pushdown automaton may go over in one move from one configuration into a finite number of new configurations. Namely, let z, q, and a be the topmost symbol in the pushdown store, the current state of the finite control, and the next input symbol, respectively. If

$$M(z, q, a) = \{(w_1, p_1), \ldots, (w_m, p_m)\}$$

with $w_i \in Z^*$ and $p_i \in K$, then the actual configuration $Wq = Qzq$ will be

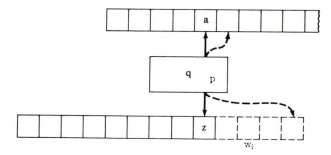

Figure 6.4 A move of a pushdown automaton.

changed into $Qw_i p_i$ (for $i = 1, \ldots, m$) and the input tape will be advanced by one square (see Figure 6.4). If for the given z and q the set $M(z, q, \lambda)$ is not empty, say,

$$M(z, q, \lambda) = \{(u_1, r_1), \ldots, (u_n, r_n)\}$$

with $u_j \in Z^*$, $r_j \in K$, then the configuration Qzq is also changed into $Qu_j r_j$ (for $j = 1, \ldots, n$) without advancing the input tape. Such moves are called λ-moves which make it possible for the pushdown automaton to change its configuration without reading any input. In a nondeterministic pushdown automaton it is, of course, possible that neither $M(z, q, a)$ nor $M(z, q, \lambda)$ is empty.

The number of possible new configurations may thus increase in each move so the best way to keep track of them all is to set up a new copy of the automaton together with its input tape separately for each individual configuration arising from each move. Each copy will then be considered to have its own life, independent of the others.

Note that any of the w_i's and u_j's above may be λ which means that the symbol z is popped from the top of the stack in that case and thus, the symbol just below it becomes the topmost symbol of the store.

The initial configuration of the pushdown automaton is always $z_0 q_0$. Let us write some word $P \in T^*$ on the input tape and start the pushdown automaton from the initial configuration $z_0 q_0$ with the read head scanning the first letter of P. Now if there is a sequence of moves which leads to a configuration Wp with $p \in H$ while reading the input word P then we say that P is accepted by the pushdown automaton. For a precise definition, first we adjust the notion of reduction.

Definition 6.7 A pushdown automaton A reduces some word $X \in Z^*KT^*$ to a word $Y \in Z^*KT^*$ in one move (in symbols, $X \underset{A}{\Rightarrow} Y$) iff there are

$z \in Z$, q, $p \in K$, $a \in T \cup \{\lambda\}$, W, $u \in Z^*$, and $P \in T^*$ such that $(u, p) \in M(z, q, a)$ with $X = WzqaP$ and $Y = WupP$.

Definition 6.8 A pushdown automaton A reduces $X \in Z^*KT^*$ to $Y \in Z^*KT^*$ (in symbols, $X \underset{A}{\Rightarrow} Y$) iff $X = Y$ or else there is a finite sequence of words $X_1, \ldots, X_n \in Z^*KT^*$ such that $X = X_1$, $Y = X_n$, and $X_i \underset{A}{\Rightarrow} X_{i+1}$ for $i = 1, \ldots, n - 1$.

Definition 6.9 The language accepted by a pushdown automaton A is defined as

$$L(A) = \left\{ P \in T^* | z_0 q_0 P \underset{A}{\overset{*}{\Rightarrow}} Wp \text{ for some } W \in Z^*, p \in H \right\}$$

Note that we require here only the existence of at least one sequence of moves leading to an accepting configuration, while several others may lead to nonaccepting ones.

The mapping M can be given again in terms of rewriting rules this way:

1) $zqa \to up \in M$ iff $(u, p) \in M(z, q, a)$
2) $zq \to up \in M$ iff $(u, p) \in M(z, q, \lambda)$

Example 6.4 Define a pushdown automaton A such that

$$L(A) = \{a^n b^n | n = 1, 2, \ldots\}$$

An intuitively simple solution is to store every a into the pushdown stack until the first b is encountered. Then each time a b is read from the input tape one a will be popped from the top of the stack. At the end of the input word we must get an empty stack whenever the input word has form $a^n b^n$. The formal definition of the pushdown automaton will be the following:

$$A = (\{z_0, a\}, \{q_0, q_1, q_2\}, \{a, b\}, M, z_0, q_0, \{q_2\})$$

where M is defined as shown in Table 6.4.

Table 6.4

M	a	b	λ
(z_0, q_0)	$(z_0 a, q_0)$	$\varnothing$	$\varnothing$
(a, q_0)	(aa, q_0)	(λ, q_1)	$\varnothing$
(z_0, q_1)	$\varnothing$	$\varnothing$	(λ, q_2)
(a, q_1)	$\varnothing$	(λ, q_1)	$\varnothing$
(z_0, q_2)	$\varnothing$	$\varnothing$	$\varnothing$
(a, q_2)	$\varnothing$	$\varnothing$	$\varnothing$

Definition 6.10 A pushdown automaton $A = (Z, K, T, M, z_0, q_0, H)$ is said to be deterministic iff for every pair $(z, q) \in Z \times K$ either (1) $M(z, q, a)$ contains exactly one element for each $a \in T$, while $M(z, q, \lambda) = \varnothing$ or (2) $M(z, q, \lambda)$ contains exactly one element while $M(z, q, a) = \varnothing$ for each $a \in T$.

According to this definition the pushdown automaton of Example 6.4 is not deterministic. But, we can make it deterministic by introducing a new state q_3, with $M(z_0, q_3, \lambda) = (z_0, q_3)$ and $M(a, q_3, \lambda) = (a, q_3)$, and replacing the empty set, $\varnothing$, by (z_0, q_3) in column a and column b of Table 6.4.

This change would not affect the accepted language but the new pushdown automaton may cycle forever without reading any further input. As a matter of fact, many pushdown automata can have infinite cycles due to the presence of λ-moves. Nevertheless, deterministic pushdown automata are strictly less powerful than nondeterministic ones. We do not prove this here, but we can make it clearer with the aid of these two languages:

$$L_1 = \{ PcP^{-1} | P \in \{a, b\}^* \}$$
$$L_2 = \{ PP^{-1} | P \in \{a, b\}^* \}$$

Both of these languages can be accepted by nondeterministic pushdown automata, but only the first one can be accepted by a deterministic pushdown automaton. The reason for that is the presence of c, which marks the center of the input string. (P can be stored in the stack and matched with P^{-1} after having found the center.) In case of L_2 we are left to our guesses when to stop pushing more symbols into the stack and start matching them with the rest of the input. Nondeterminism allows us to guess both ways all the time, which will insure that we also include the correct guess among the others.

Our next aim is to show that the languages accepted by nondeterministic pushdown automata are exactly the context-free languages. To this end we define the null-language of A, $N(A)$, that is accepted by empty store.

Definition 6.11 The language $N(A)$ accepted by a pushdown automaton A with empty store is defined as

$$N(A) = \left\{ P \in T^* | z_0 q_0 P \overset{*}{\underset{A}{\Rightarrow}} p \text{ for some } p \in K \right\}$$

Note that if the store is empty, then the automaton is blocked, since no move is defined for an empty store. (This is why we need z_0 for the initial configuration.) The set of accepting states is irrelevant for $N(A)$.

LEMMA C For every pushdown automaton A we can give a pushdown automaton A' such that $N(A') = L(A)$.

PROOF Let $A = (Z, K, T, M, z_0, q_0, H)$ be a pushdown automaton. We construct A' such that it simulates the moves of A, but whenever A reaches an accepting state, then A' has the option of clearing the entire pushdown store. A new initial symbol $z_0' \neq z_0$ is needed to prevent A' from accepting some $P \in L(A)$ which empties the stack in A. So let A' be defined as

$$A' = (Z \cup \{z_0'\}, K \cup \{q_0', q_h'\}, M', z_0', q_0', \varnothing)$$

where $z_0' \notin Z$, $q_0', q_h' \notin K$, and M' are defined as follows:

1) $M'(z_0', q_0', \lambda) = \{(z_0' z_0, q_0)\}$
2) $M'(z, q, a) = M(z, q, a)$ for every $z \in Z$, $q \in K$, and $a \in T$
3) $M(z, q, \lambda) \subseteq M'(z, q, \lambda)$ for every $z \in Z$, and $q \in K$
4) $(\lambda, q_h') \in M'(z, q, \lambda)$ for every $z \in Z \cup \{z_0'\}$, and $q \in H \cup \{q_h'\}$

Now, assume that $P \in L(A)$, that is, $z_0 q_0 P \underset{A}{\overset{*}{\Rightarrow}} Wq$ for some $W \in Z^*$ and $q \in H$; then we obviously have

$$z_0' q_0' P \underset{A'}{\Rightarrow} z_0' z_0 q_0 P \underset{A'}{\overset{*}{\Rightarrow}} z_0' Wq \underset{A'}{\overset{*}{\Rightarrow}} q_h'$$

which means that $P \in N(A')$.

Conversely, let $P \in N(A')$, that is, $z_0' q_0' P \underset{A'}{\overset{*}{\Rightarrow}} q$ for some $q \in K \cup \{q_0', q_h'\}$. The first step here can only be $z_0' q_0' P \underset{A'}{\Rightarrow} z_0' z_0 q_0 P$ since no other move is defined for (z_0', q_0'). Further, the symbol z_0' cannot be cleared from the bottom of the stack unless we use a λ-move from group 4. Hence, there is some $q \in H$ and $W \in Z^*$ such that

$$z_0' z_0 q_0 P \underset{A'}{\overset{*}{\Rightarrow}} z_0' Wq \underset{A'}{\overset{*}{\Rightarrow}} q_h'$$

where

$$z_0 q_0 P \underset{A}{\overset{*}{\Rightarrow}} Wq$$

which completes the proof.

Theorem 6.5 For every context-free grammar G, we can give a pushdown automaton A such that $L(A) = L(G)$.

PROOF We can assume that $G = (V_N, V_T, S, F)$ is in Chomsky normal form except possibly for the rule $S \rightarrow \lambda$, if $\lambda \in L(G)$, but then S does not occur on the right-hand sides of the rules.

Define the pushdown automaton

$$A = (V_N \cup \{z_0\}, K, V_T, M, z_0, q_0, \{q_h\})$$

such that

$$K = \left(\bigcup_{X \in V_N} q_X \right) \cup \{q_0, q_h\}$$

and M is the following:

1) $z_0 q_0 \to z_0 q_S \in M$ iff $S \to \lambda \in F$
2) $z_0 q_0 a \to z_0 q_X \in M$ iff $X \to a \in F$
3) $Z q_Y a \to Z Y q_X \in M$ for every $Z \in V_N \cup \{z_0\}$ and $Y \in V_N$ iff $X \to a \in F$
4) $Z q_Y \to q_X \in M$ iff $X \to Z Y$
5) $z_0 q_S \to q_h \in M$

Note that these moves are essentially the same as the inverted rules of G.

To prove $L(A) \subseteq L(G)$, we assume that $P \in L(A)$, that is, $z_0 q_0 P \overset{*}{\underset{A}{\Rightarrow}} W q_h$ for some $W \in Z^*$. But, because of the construction of A, here we must have $W = \lambda$ or more exactly

$$z_0 q_0 P \overset{*}{\underset{A}{\Rightarrow}} z_0 q_S \underset{A}{\Rightarrow} q_h$$

If $P = \lambda$ here, then the first step is $z_0 q_0 \Rightarrow z_0 q_S$ and thus $S \to \lambda \in F$. Otherwise we have

$$z_0 q_0 P \underset{A}{\Rightarrow} z_0 q_X Q \overset{*}{\underset{A}{\Rightarrow}} z_0 q_S \underset{A}{\Rightarrow} q_h$$

where $P = aQ$ and $X \to a \in F$ for some $a \in V_T$. Now it is easy to show that

$$z_0 q_0 P \overset{*}{\underset{A}{\Rightarrow}} z_0 q_S$$

implies

$$S \overset{*}{\underset{G}{\Rightarrow}} P$$

Namely, to each move of type 2, 3, or 4 in M we can find the corresponding rule in F so we can construct the corresponding derivation in G by following the above reduction in the reverse sense.

Conversely, if $P \in L(G)$, that is, we have a derivation $S \overset{*}{\underset{G}{\Rightarrow}} P$ then we can construct the reduction $z_0 q_0 P \overset{*}{\underset{A}{\Rightarrow}} z_0 q_S$, which gives $P \in L(A)$, and this completes the proof.

Example 6.5 Develop a pushdown automaton that accepts the language generated by the grammar given in Example 3.1.

In Example 3.1 we have converted this grammar into Chomsky normal form. All we have to do now is to use the construction of Theorem 6.5. So we get

$$A = (V_N \cup \{z_0\}, K, V_T, M, z_0, q_0, \{q_h\})$$

where

$$V_T = \{a, b, c, +, (,)\}$$
$$V_N = \{S, A, B, C, D, E, X, Y\}$$

and therefore

$$K = \{q_S, q_A, q_B, q_C, q_D, q_E, q_X, q_Y, q_0, q_h\}$$

The moves in M will be defined this way

1) This group is empty because $S \to \lambda \notin F$.

2) $z_0 q_0 a \to z_0 q_S \qquad z_0 q_0 b \to z_0 q_S$

   $z_0 q_0 a \to z_0 q_A \qquad z_0 q_0 b \to z_0 q_A$

   $z_0 q_0 a \to z_0 q_B \qquad z_0 q_0 b \to z_0 q_B$

   $z_0 q_0 c \to z_0 q_S \qquad z_0 q_0 + \to z_0 q_C$

   $z_0 q_0 c \to z_0 q_A \qquad z_0 q_0 (\to z_0 q_D$

   $z_0 q_0 c \to z_0 q_B \qquad z_0 q_0) \to z_0 q_E$

3) This group contains a very large number of moves like

   $A q_B a \to A B q_S$

   $A q_B a \to A B q_A$

   $A q_B a \to A B q_B \quad$ and so forth

4) $S q_X \to q_S \qquad S q_E \to q_Y$

   $A q_B \to q_A \qquad A q_B \to q_S$

   $D q_Y \to q_B \qquad D q_Y \to q_A$

   $C q_A \to q_X \qquad D q_Y \to q_S$

5) $z_0 q_S \to q_h$

    Note that the large number of the moves in groups 3 is due to the fact that they depend only on the input symbol and not on the current state or on the symbol on the top of the stack.

    In order to prove the converse of Theorem 6.5 we have to apply a different approach, because in general we would not get a context-free grammar simply by inverting the moves of a pushdown automaton. Even worse, by inverting the rules of form $zqa \rightarrow wp$ or $zq \rightarrow wp$ we may get length-increasing, as well as length-decreasing, rules.

A derivation of some terminal word P can be constructed, however, basically in two different ways:

1) *Bottom-up construction.* This means that we start with P and try to reduce it to S by using the rules of the grammar in the reverse sense.
2) *Top-down construction.* Here we start with S and try to derive terminal strings to be matched with P.

In the proof of the following theorem we shall adopt the second approach. Each move of the pushdown automaton will thus be considered as a tentative step in constructing the derivation $S \overset{*}{\underset{G}{\Rightarrow}} P$, top-down, by guessing the rule to be applied next. The nondeterministic operation of the automaton permits the simultaneous construction of every possible derivation that may eventually lead to the given P.

Theorem 6.6 For every pushdown automaton A we can give a context-free grammar G such that $L(G) = N(A)$.

PROOF Given a pushdown automaton $A = (Z, K, T, M, z_0, q_0, \varnothing)$ we define the context-free grammar $G = (V_N, V_T, S, F)$ such that $V_T = T$ and the elements of V_N are the ordered triples of form $[q, x, p]$, where $p, q \in K$ and $x \in Z$. Further, we use a new symbol for S, that is, we have $V_N = (K \times Z \times K) \cup \{S\}$. The rules in F will be defined as follows:

1) Let $S \rightarrow [q_0, z_0, p] \in F$ for all $p \in K$.
2) If $xqa \rightarrow y_1 \cdots y_m p_m \in M$ with $a \in T \cup \{\lambda\}$, then for every $p_0, p_1, \ldots, p_{m-1} \in K$ let

$$[q, x, p_0] \rightarrow a[p_m, y_m, p_{m-1}] \cdots [p_1, y_1, p_0] \in F$$

(For $m = 0$, that is, for $xqa \rightarrow p \in M$, let $[q, x, p] \rightarrow a \in F$.)
3) No other rules are in F.

First we show that $L(G) \subseteq N(A)$. To this end we prove that for every $x \in Z, p, q \in K$, and $P \in T^*$ the relation

$$[q, x, p] \overset{*}{\underset{G}{\Rightarrow}} P \text{ implies } xqP \overset{*}{\underset{A}{\Rightarrow}} p$$

We use induction on the number of steps in the derivation $[q, x, p] \overset{*}{\underset{G}{\Rightarrow}} P$. For one step we have obviously $P = a \in T \cup \{\lambda\}$ and $xqa \rightarrow p \in M$. Assume that the assertion is valid for every derivation with at most n steps, and let the derivation $[q, x, p] \overset{*}{\underset{G}{\Rightarrow}} P$ have $n + 1$ steps. Then it must be of form

$$[q, x, p_0] \underset{G}{\Rightarrow} a[p_m, y_m, p_{m-1}] \cdots [p_1, y_1, p_0] \overset{*}{\underset{G}{\Rightarrow}} P$$

Hence, there are words $P_m, P_{m-1}, \ldots, P_1 \in T^*$ such that $P = aP_m P_{m-1} \cdots P_1$ and

$$[p_i, y_i, p_{i-1}] \overset{*}{\underset{G}{\Rightarrow}} P_i \text{ for } i = 1, \ldots, m$$

Then by the induction hypothesis

$$y_i p_i P_i \overset{*}{\underset{A}{\Rightarrow}} p_{i-1}$$

and by the definition of G

$$xqa \underset{A}{\Rightarrow} y_1 \cdots y_m P_m$$

Hence, we get

$$xqP = xqaP_m \cdots P_1 \underset{A}{\Rightarrow} y_1 \cdots y_m p_m P_m \cdots P_1$$

$$\overset{*}{\underset{A}{\Rightarrow}} y_1 \cdots y_{m-1} P_{m-1} P_{m-1} \cdots P_1 \overset{*}{\underset{A}{\Rightarrow}} p_0 = p$$

which was to be shown.

Now, if $P \in L(G)$, then there is some $p \in K$ such that

$$S \underset{G}{\Rightarrow} [q_0, z_0, p] \overset{*}{\underset{G}{\Rightarrow}} P$$

and thus

$$z_0 q_0 P \overset{*}{\underset{A}{\Rightarrow}} p$$

which means that $P \in N(A)$.

For the reverse inclusion we show that the relation

$$xqP \overset{*}{\underset{A}{\Rightarrow}} p \quad \text{implies} \quad [q, x, p] \overset{*}{\underset{G}{\Rightarrow}} P$$

We use induction on the number of moves involved in the given reduction. For a single move we have $P = a$ and $xqa \to p \in M$ which gives $[q, x, p] \to a \in F$. For more than one move $xqP \overset{*}{\underset{A}{\Rightarrow}} p$ must have the form

$$xqP = xqaQ \underset{A}{\Rightarrow} y_1 \cdots y_m p_m Q \overset{*}{\underset{A}{\Rightarrow}} p$$

Let us follow this reduction after the first move until the symbol y_{m-1} becomes the top of the stack. This means that there is some $P_m \in T^*$ such that $Q = P_m Q_1$ and $y_m p_m P_m \underset{A}{\overset{*}{\Rightarrow}} p_{m-1}$ for some $p_{m-1} \in K$, and thus

$$xqP \underset{A}{\Rightarrow} y_1 \cdots y_m p_m P_m Q_1 \underset{A}{\overset{*}{\Rightarrow}} y_1 \cdots y_{m-1} p_{m-1} Q_1$$

By continuing this reasoning we get some $P_{m-1}, \ldots, P_1$ such that

$$P = a P_m P_{m-1} \cdots P_1$$

and

$$y_i p_i P_i \underset{A}{\overset{*}{\Rightarrow}} p_{i-1} \quad \text{for } i = 1, \ldots, m$$

By the induction hypothesis we get

$$[p_i, y_i, p_{i-1}] \underset{G}{\overset{*}{\Rightarrow}} P_i \quad \text{for } i = 1, \ldots, m$$

and by the definition of G

$$[q, x, p] \underset{G}{\Rightarrow} a[p_m, y_m, p_{m-1}] \cdots [p_1, y_1, p]$$

which gives the result $[q, x, p] \underset{G}{\overset{*}{\Rightarrow}} P$.

Now, if $P \in N(A)$ then $z_0 q_0 P \underset{A}{\overset{*}{\Rightarrow}} p$ for some $p \in K$ and thus

$$S \underset{G}{\Rightarrow} [q_0, z_0, p] \underset{G}{\overset{*}{\Rightarrow}} P$$

which means that $N(A) \subseteq L(G)$, and this completes the proof.

Corollary 1 For every pushdown automaton A, we can give a pushdown automaton A' such that $L(A') = N(A)$.

PROOF This follows from Theorems 6.6 and 6.5.

Corollary 2 For every pushdown automaton A, we can give a context-free grammar G such that $L(G) = L(A)$.

PROOF The assertion follows from Lemma C and Theorem 6.6.

REMARK Corollary 1 can also be shown directly with the aid of a construction very similar to the one we have used in Lemma C. This means that the two different ways of accepting languages by pushdown automata are essentially the same and equivalent to the context-free grammars.

Note that the rules in group 2 are in Greibach normal form, except for those rules which correspond to a λ-move.

EXERCISES

6.10 Develop a pushdown automaton that accepts the language given in Exercise 1.2b).

6.11 Develop a pushdown automaton that accepts the language given in Example 1.1. Find also an equivalent deterministic pushdown automaton.

6.12 Develop a pushdown automaton that accepts the language $\langle a^n b^n c^k \mid n \geqslant 1,\ k \geqslant 1 \rangle$ given in Example 3.3.

6.13 Design a deterministic pushdown automaton to accept with empty store the language

$$\langle a^k b^l c^m d^n \mid k, l, m, n \geqslant 0 \text{ and } m = k + l + n \rangle$$

6.14 Design a nondeterministic pushdown automaton to accept the set of regular expressions over the alphabet $\langle a, b, c \rangle$.

6.15 Find a context-free grammar to generate the language accepted by the following pushdown automaton with empty store:

$$A = (\langle z_0, x \rangle, \langle q_0, q_1 \rangle, \langle a, b, c \rangle, M, z_0, q_0, \varnothing)$$

where the transition function M contains the moves

$$z_0 q_0 a \rightarrow x q_0 \qquad z_0 q_0 b \rightarrow x q_0$$
$$x q_0 a \rightarrow x x q_0 \qquad x q_0 b \rightarrow x x q_0$$
$$x q_0 c \rightarrow q_1 \qquad x q_1 c \rightarrow q_1$$

Note that this pushdown automaton is nondeterministic and it may get stuck with some inputs.

6.3 TWO-PUSHDOWN AUTOMATA

The first idea to increase the power of pushdown automata is to attach one more pushdown store to the finite state control unit. Interestingly enough, two-pushdown automata are already so powerful that their power cannot be increased any further (this will be dealt with in greater detail in Chapter 7). In the present section we shall prove that the languages accepted by two-pushdown automata are exactly the type 0 languages. Then what can we say about type 1 languages? We shall see that they can be recognized by a restricted type of two-pushdown automata called a *linear bounded automaton* where the size of the working space, i.e., the amount of storage space used for the recognition process is limited by the size of the input. Such a characterization may appear strange at first since we have not made any restriction on the size of the stores in the case of a single pushdown store. This means that a single pushdown store with arbitrary size is strictly less powerful than two pushdown stores with limited size. The fact is that the size of the pushdown store of a pushdown automaton need not grow arbitrarily either, provided that the input word is accepted.

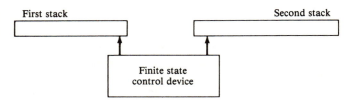

Figure 6.5 Two-pushdown automaton.

Definition 6.12 A two-pushdown automaton is an ordered seventuple

$$A = (Z, K, T, M, z_0, q_0, H)$$

where Z is a finite set of *pushdown symbols*
K is a finite set of *internal states*
$T \subseteq Z$ is a finite set of *input symbols*
M is a mapping from $Z \times K \times Z$ into the subsets of

$$Z \times K \times \{R, L, N, E, I\}, \text{ the } \textit{transition function}$$

$z_0 \in Z - T$ is the *initial symbol*
$q_0 \in K$ is the *initial state*
$H \subseteq K$ is the set of *accepting states*

A two-pushdown automaton is represented in Figure 6.5, where the top-most symbol in each store is scanned by the corresponding read-write head. Initially, the first stack is empty, i.e., it contains only the initial symbol z_0 while the input word is written in the second stack in the reverse order such that its last symbol is placed at the bottom and the first symbol is placed at the top of the stack. Note that we have here no extra input tape so the input word may be overwritten in the stack in the course of the recognition procedure.

Definition 6.13 A configuration of a two-pushdown automaton is a word of the form UqW, where $U, W \in Z^*$ and $q \in K$. (Here U and W are representing the contents of the two pushdown stores, q is the actual internal state, and the two read-write heads are scanning the last symbol of U and the first symbol of W, respectively.)

The transition function M defines the moves of the two-pushdown automaton which depend on the internal state and on the two symbols scanned by the read-write heads. There are five different types of moves denoted by R, L, N, E, and I, which represent right-shift, left-shift, no-shift, erase, and insert, respectively. Each move involves the printing of a symbol on the top of one of the two stacks. The way the configuration is changed by these moves can be best described in terms of rewriting rules as follows:

1) $xqy \rightarrow xzp \in M$ iff $(z, p, R) \in M(x, q, y)$
2) $xqy \rightarrow pzy \in M$ iff $(z, p, L) \in M(x, q, y)$
3) $xqy \rightarrow xpz \in M$ iff $(z, p, N) \in M(x, q, y)$
4) $xqy \rightarrow pz \in M$ iff $(z, p, E) \in M(x, q, y)$
5) $xqy \rightarrow xpzy \in M$ iff $(z, p, I) \in M(x, q, y)$

The transition function M is multivalued in general, so we can have several rewriting rules in M with the same left-hand side. The reduction relations $\underset{A}{\Rightarrow}$ and $\underset{A}{\overset{*}{\Rightarrow}}$ are defined in the usual way on the basis of the rules in M. The configuration of a nondeterministic two-pushdown automaton may go over to a finite number of new configurations in one move. Therefore, we have to set up a new copy of the automaton together with its pushdown stores for each different configuration resulting from one move, and each of these configurations will then be treated independently of the others. The number of configurations can thus increase very rapidly with the number of moves.

Definition 6.14 The language accepted by a two-pushdown automaton is defined as

$$L(A) = \left\{ P \in T^* | z_0 q_0 P \overset{*}{\underset{A}{\Rightarrow}} Up \text{ for some } U \in Z^+ \text{ and } p \in H \right\}$$

Definition 6.15 The language accepted by a two-pushdown automaton with empty store is defined as

$$N(A) = \left\{ P \in T^* | z_0 q_0 P \overset{*}{\underset{A}{\Rightarrow}} pz \text{ for some } p \in K \text{ and } z \in Z \right\}$$

Note that the empty word λ is in $L(A)$ iff $q_0 \in H$. The definition of $N(A)$, however, implies that $\lambda \notin N(A)$ for every A because z_0 cannot be removed from the first stack without making any move. We shall use, therefore, the convention $q_0 \in H$ also for $N(A)$ to include the empty word in the language.

Theorem 6.7 For every two-pushdown automaton A we can give a type 0 grammar G such that $L(G) = N(A)$.

PROOF Let $A = (Z, K, T, M, z_0, q_0, H)$ be a two-pushdown automaton. We define the grammar $G = (V_N, V_T, S, F)$ such that

$$V_N = (Z - T) \cup (Z \times K \times Z) \cup \{S\}$$

where $S \notin Z$, $V_T = T$, and the rules in F are as follows:

1) $x[z, p, u] \rightarrow [x, q, y]u \in F$ for all $u \in Z$ iff $xqy \rightarrow xzp \in M$
2) $[u, p, z]y \rightarrow u[x, q, y] \in F$ for all $u \in Z$ iff $xqy \rightarrow pzy \in M$

3) $[x, p, z] \rightarrow [x, q, y] \in F$ iff $xqy \rightarrow xpz \in M$

4) $[u, p, z] \rightarrow u[x, q, y] \in F$ for all $u \in Z$ iff $xqy \rightarrow pz \in M$

5) $[x, p, z]y \rightarrow [x, q, y] \in F$ iff $xqy \rightarrow xpzy \in M$

6) $[z_0, q_0, x] \rightarrow x \in F$ for all $x \in T$

7) $S \rightarrow [z_0, q, y] \in F$ iff $z_0qy \rightarrow pz \in M$ for some $p \in K$ and $z \in Z$

8) $S \rightarrow \lambda \in F$ iff $q_0 \in H$

As can be seen from the construction, the rules of G are, in fact, the inverted moves of A. The brackets form only windows through which we combine three symbols into one.

It is clear that $\lambda \in L(G)$ iff $\lambda \in N(A)$. Assume that $P \in N(A)$ for some $P \in T^+$. Then by definition $z_0q_0P \overset{*}{\underset{A}{\Rightarrow}} pz$ where the last step is of form $z_0qy \rightarrow pz$, and thus $S \underset{G}{\Rightarrow} [z_0, q, y]$. Further, to each move in the reduction $z_0q_0P \overset{*}{\underset{A}{\Rightarrow}} z_0qy$ there is a rule in F which corresponds to the inverse of that move. So we can construct the derivation

$$[z_0, q, y] \overset{*}{\underset{G}{\Rightarrow}} [z_0, q_0, x]P_1$$

where $P = xP_1$. Finally we can apply a rule from group 6 which yields $S \overset{*}{\underset{G}{\Rightarrow}} P$.

Conversely, let $P \in L(G)$ and $P \in T^+$. Then we have a derivation $S \overset{*}{\underset{G}{\Rightarrow}} P$ where the first step must be of the form $S \underset{G}{\Rightarrow} [z_0, q, y]$. Now for each step in the rest of the derivation we can find a corresponding reverse move until we reach the last step of the derivation. The latter must be the application of a rule from group 6. But then it must be applied at the first position of the word, otherwise the initial z_0 cannot disappear. Hence, we have

$$S \underset{G}{\Rightarrow} [z_0, q, y] \overset{*}{\underset{G}{\Rightarrow}} [z_0, q_0, x]P_1 \underset{G}{\Rightarrow} xP_1 = P$$

from which the reduction

$$z_0q_0P \overset{*}{\underset{A}{\Rightarrow}} z_0qy \underset{A}{\Rightarrow} pz$$

can be easily constructed which means that $L(G) \subseteq N(A)$, and this completes the proof.

Corollary If the two-pushdown automaton A does not contain any inserting moves (that is, $xqy \rightarrow xpzy$ type moves) then the language $N(A)$ is of type 1.

PROOF This follows immediately from the construction of G in the above proof because the rules in F are all length-increasing (nondecreasing) except for group 5.

Definition 6.16 Two-pushdown automata having no inserting moves are called *linear bounded automata*.

If a linear bounded automaton is started from the initial configuration $z_0 q_0 P$, then for every configuration UqW that occurs during the operation of the automaton we have

$$|UqW| \leqslant |z_0 q_0 P|$$

Theorem 6.8 For every type 0 grammar G there is a two-pushdown automaton A such that $L(A) = L(G)$.

PROOF We can assume that the grammar G is in the normal form given in Theorem 5.1. Let us define the two-pushdown automaton A as follows

$$A = (V_N \cup V_T \cup \{z_0\}, \{q_0, q_1, q_2\}, V_T, M, q_0, z_0, H)$$

where $z_0 \notin V_N \cup V_T$ and $H = \{q_2\}$, if $\lambda \notin L(G)$ and $H = \{q_0, q_2\}$ if $\lambda \in L(G)$. The moves in M will be the following:

1) $z_0 q_0 a \rightarrow z_0 q_1 z \in M$ iff $z \rightarrow a \in F, a \in V_T$
2) $x q_1 y \rightarrow x q_1 z \in M$ for all $x \in V_N \cup \{z_0\}$ iff $z \rightarrow y \in F$
3) $x q_1 y \rightarrow q_1 z \in M$ iff $z \rightarrow xy \in F$
4) $x q_1 y \rightarrow x q_1 z \in M$ iff $xz \rightarrow xy \in F$
5) $x q_1 y \rightarrow q_1 zy \in M$ iff $zy \rightarrow xy \in F$
6) $x q_1 y \rightarrow x q_1 zy \in M$ for all $x \in V_N \cup \{z_0\}$ iff $zy \rightarrow y \in F$
7) $x q_1 y \rightarrow q_1 xy \in M$ for all $x, y \in V_N$
8) $x q_1 y \rightarrow xy q_1 \in M$ for all $x \in V_N \cup \{z_0\}$ and $y \in V_N$
9) $z_0 q_1 S \rightarrow z_0 S q_2 \in M$

Again it is clear that $\lambda \in L(A)$ iff $\lambda \in L(G)$. To prove $L(G) \subseteq L(A)$, assume that $P \in L(G)$ and $P \neq \lambda$. The order of the applications of the rules in the derivation $S \overset{*}{\underset{G}{\Rightarrow}} P$ can be chosen in such a way that the last step is the application of a rule $z \rightarrow a$ at the first position of P.

$$S \overset{*}{\underset{G}{\Rightarrow}} zP_1 \underset{G}{\Rightarrow} aP_1 = P$$

Hence, by move 1 we get

$$z_0 q_0 P = z_0 q_0 a P_1 \underset{A}{\Rightarrow} z_0 q_1 z P_1$$

According to moves 2, 3, 4, 5, and 6 for each step in the derivation

$$S \overset{*}{\underset{G}{\Rightarrow}} zP_1$$

we can find a reverse move with q_1 inserted in the appropriate position. But by moves 7 and 8 the position of q_1 can be shifted arbitrarily within the word to be reduced. Therefore, we can find a sequence of moves that will produce the reduction

$$z_0 q_0 P \underset{A}{\Rightarrow} z_0 q_1 z P_1 \overset{*}{\underset{A}{\Rightarrow}} z_0 q_1 S$$

and by move 9 we get $P \in L(A)$.

Conversely, if $z_0 q_0 P \overset{*}{\underset{A}{\Rightarrow}} U q_2$ then $U = z_0 S$ must be the case because q_2 occurs only in the move given by 9. Therefore, also $z_0 q_0 P \overset{*}{\underset{A}{\Rightarrow}} z_0 q_1 S$ holds and the derivation $S \overset{*}{\underset{G}{\Rightarrow}} P$ can be constructed in the same manner as before. So we get $L(A) \subseteq L(G)$, and this completes the proof.

Corollary 1 Each type 1 language is accepted by a linear bounded automaton.

PROOF If the grammar is length-increasing then group 6 in the above construction will be empty, and thus the automaton will have no inserting moves.

Corollary 2 For every two-pushdown automaton A, there is a two-pushdown automaton A' such that $L(A') = N(A)$.

PROOF This follows from Theorems 6.7 and 6.8. The converse of Corollary 2 will be shown directly.

Theorem 6.9 For every two-pushdown automaton A, we can give a two-pushdown automaton A' such that $N(A') = L(A)$.

PROOF Given a two-pushdown automaton $A = (Z, K, T, z_0, q_0, H)$, we define A' such that

$$A' = (Z \cup \{z_0'\}, K \cup \{q_1', q_2'\}, T, M', z_0', q_0, H')$$

where $z_0' \notin Z$, $q_1', q_2' \notin K$, $H' = \varnothing$ if $\lambda \notin L(A)$, $H' = \{q_0\}$ if $\lambda \in L(A)$, and M' is the following:

1) $M(x, q, y) \subseteq M'(x, q, y)$ for all $x, y \in Z$ and $q \in K$
2) $(z, p, R) \in M'(z_0', q, y)$ iff $(z, p, R) \in M(z_0, q, y)$
3) $(z, p, N) \in M'(z_0', q, y)$ iff $(z, p, N) \in M(z_0, q, y)$

4) $(z, p, I) \in M'(z_0', q, y)$ iff $(z, p, I) \in M(z_0, q, y)$

5) $(z, q_1', N) \in M'(z_0', q, y)$ iff $(z, p, L) \in M(z_0, q, y)$ or $(z, p, E) \in M(z_0, q, y)$

6) $(z, q_2', E) \in M'(x, q, y)$ iff $(z, p, R) \in M(x, q, y)$ and $p \in H$

7) $(y, q_2', E) \in M'(x, q_2', y)$ for all $x \in Z \cup \{z_0'\}$ and $y \in Z$

As can be seen from this construction the automaton A' started from the initial configuration $z_0' q_0 P$ would simulate the moves of A as started from $z_0 q_0 P$ (z_0' takes the role of z_0) until in A either (1) we get a configuration where the first stack is empty or (2) the finite control goes to a state $p \in H$ in a right-shift move.

In the first case A' goes to state q_1' which halts the operation of A'. This means that A' gets stuck whenever the first stack in A becomes empty.

In the second case, according to 6, A' would go also to state q_2' in an erase move besides the right-shift. Then by 7 it can completely erase its first stack while leaving the contents of the second stack unchanged. Hence, the relation

$$z_0 q_0 P \overset{*}{\underset{A}{\Rightarrow}} Wp$$

with some $p \in H$ and $W \in Z^+$ implies that

$$z_0' q_0 P \overset{*}{\underset{A'}{\Rightarrow}} q_2' y$$

for some $y \in Z$. So we have $L(A) \subseteq N(A')$.

Conversely, if $P \in N(A')$ that is $z_0' q_0 P \overset{*}{\underset{A'}{\Rightarrow}} pz$ then we must have here $p = q_2'$ since z_0' could not be erased otherwise. But by 6, q_2' can appear in a reduction only for some $p \in H$. Therefore, we have $z_0' q_0 P \overset{*}{\underset{A'}{\Rightarrow}} z_0' Wp$ for some $W \in Z^*$ and $p \in H$ which implies that $z_0 q_0 P \overset{*}{\underset{A}{\Rightarrow}} z_0 Wp$ and thus, $P \in L(A)$, which completes the proof.

This way we have established the equivalence between type 0 grammars and two-pushdown automata no matter which of the two definitions is used for accepting a language. At the same time we have seen that type 1 grammars are equivalent to linear bounded automata as defined in Definition 6.16.

In Chapter 8 we shall prove that nondeterministic two-pushdown automata are equivalent to the deterministic ones. For linear bounded automata, however, it is not known whether the nondeterministic version is strictly more powerful than the deterministic one. This is the so called LBA problem, which is probably the most famous unsolved problem in the theory of automata.

Example 6.6 Develop a linear bounded automaton that accepts the language $\{a^n b^n c^n \mid n \geqslant 1\}$.

For the construction of the linear bounded automaton we make use of the grammar given in Example 1.2. First we convert this grammar to Kuroda normal form. Three new variables A, B, and C will be assigned to the terminal letters and three more variables, E, D, and U will be introduced to reduce the length of the right-hand sides of the rules. So we get

$$
\begin{array}{lll}
S \to AD & XB \to BX & A \to a \\
S \to ED & BY \to YB & B \to b \\
E \to AX & XC \to UC & C \to c \\
D \to BC & AY \to AE & \\
U \to YD & AY \to AA &
\end{array}
$$

Next we introduce four more variables to obtain the refined version of Kuroda normal form.

$$
\begin{array}{llll}
S \to AD & S \to ED & E \to AX & D \to BC \\
U \to YD & A \to a & B \to b & C \to c \\
XB \to X'B & X'B \to X'B' & X'B' \to BB' & BB' \to BX \\
BY \to B''Y & B''Y \to B''Y' & B''Y' \to YY' & YY' \to YB \\
XC \to UC & AY \to AE & AY \to AA &
\end{array}
$$

Now we can use Theorem 6.8 for the construction of the equivalent two-pushdown automaton. The reader should complete the construction and examine the way of accepting the words abc, $aabbcc$, and $aaabbbccc$.

6.4 TURING MACHINES

In the literature on formal languages the role of our two-pushdown automata is usually played by Turing machines. The reason for that is most likely the fact that the latter were already introduced in the mid-thirties by the English mathematician A. M. Turing, who used them for the development of the theory of computability. There are several versions of the notion of Turing machines which can be shown to be equivalent to the original one. In fact, our two-pushdown automaton is just another version of the original concept.

The original definition of the Turing machine includes a finite state control device and a two-way infinite tape which is scanned by a read-write head (see Figure 6.6). The tape is divided into squares each of which can hold one symbol of a finite alphabet Z. Initially we write the input word (a finite

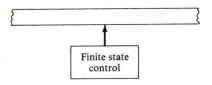

Figure 6.6 Turing machine.

sequence of symbols) on the tape while the rest of the tape is assumed to be empty. An empty square is supposed to contain the blank symbol B, which is not included in the alphabet Z. It is usually required that in each move the Turing machine should write a symbol from Z on the tape so that the symbol B can only be read but never written by the machine. This means that in the course of the operation of the machine the tape would contain a single contiguous string of nonblank symbols provided that the read-write head is placed initially to one of the squares holding the input word.

Each move of the Turing machine consists of the following actions:

The symbol currently scanned by the read-write head is read by the control device.
The current state of the control device is changed.
A symbol is written on the tape to replace the symbol just read.
The read-write head shifts its position on the tape to the right or to the left by one square.

Thus, each move can be characterized by an ordered quintuple of the form

$$(q, x, p, y, R) \quad \text{or} \quad (q, x, p, y, L)$$

where q is the current state, x is the current tape symbol, p is the new state, y is the symbol to be written, and $R(L)$ denotes the right (left) shift of the read-write head. A Turing machine can be defined as a finite set of ordered quintuples of the above form. If a Turing machine contains at most one quintuple beginning with q and x for each pair (q, x) then it is deterministic, otherwise it is nondeterministic. The formal definition will be given here in a similar fashion as the automata defined so far.

Definition 6.17 A nondeterministic Turing machine is an ordered sixtuple $A = (Z, K, T, M, q_0, B)$

where Z is a finite alphabet of *tape symbols*
   K is a finite set of *internal states*
   $T \subseteq Z$ is a finite set of *input symbols*
   M is a mapping from $K \times (Z \cup \{B\})$ into the subsets of $K \times Z \times \{R, L\}$; the *transition function*
   $q_0 \in K$ is the *initial state*
   $B \notin Z$ is the *blank symbol*

The *configuration* of a Turing machine is a word of the form UqW or qBW where $q \in K$ is the current state, U and W are in Z^*, and they are representing the nonblank portions of the tape while the read-write head is scanning the first symbol immediately to the right of q. The moves of the Turing machine define the relation $\underset{A}{\Rightarrow}$ between the configurations as follows:

1) $UqyW \underset{A}{\Rightarrow} UzpW$ iff $(p, z, R) \in M(q, y)$

2) $Uq \underset{A}{\Rightarrow} Uzp$ iff $(p, z, R) \in M(q, B)$

3) $UxqyW \underset{A}{\Rightarrow} UpxzW$ iff $(p, z, L) \in M(q, y)$

4) $Uxq \underset{A}{\Rightarrow} Upxz$ iff $(p, z, L) \in M(q, B)$

5) $qBW \underset{A}{\Rightarrow} zpW$ iff $(p, z, R) \in M(q, B)$

6) $qyW \underset{A}{\Rightarrow} pBzW$ iff $(p, z, L) \in M(q, y)$

7) $qBW \underset{A}{\Rightarrow} pBzW$ iff $(p, z, L) \in M(q, B)$

where U and W are in Z^* and x, y, z are in Z. The reflexive and transitive closure of $\underset{A}{\Rightarrow}$ is again denoted by $\underset{A}{\overset{*}{\Rightarrow}}$.

Definition 6.18 A configuration is called a *halting configuration* iff it is of the form either $UqyW$ with $y \in Z$ and $M(q, y) = \varnothing$, or else Uq or qBW with $M(q, B) = \varnothing$.

A word $P \in T^*$ is accepted by a Turing machine if and only if the Turing machine started from the initial configuration q_0P is able to reach to a halting configuration in a finite number of moves. As we are dealing with nondeterministic machines we require only the existence of at least one sequence of moves leading to a halting configuration, while at the same time there may exist any number of infinite sequences of moves. The classical definition of the Turing machine has covered only the deterministic case where in each configuration there is at most one possible move. We shall see in Chapter 8 that nondeterminism with respect to Turing machines is just a technical matter with no influence on the computing power. By the way, a sequence of moves of a Turing machine is called a computation as it is meant to compute some result from the input.

Definition 6.19 The language accepted by a Turing machine is the set

$$L(A) = \left\{ P \in T^* \mid q_0P \underset{A}{\overset{*}{\Rightarrow}} X \text{ for some halting configuration } X \right\}$$

Intuitively it should be clear that Turing machines and two-pushdown automata are fairly similar. The tape of the Turing machine can be imagined as

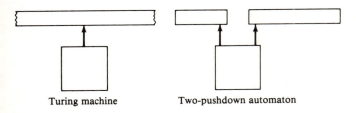

Turing machine   Two-pushdown automaton

Figure 6.7

being cut up at the read-write head and the nonblank portions of the two parts can be placed into two stacks (see Figure 6.7). A right or left shift of the original tape would correspond to popping the top of one stack while adding a new symbol to the other. Extending the nonblank portion of the original tape would be represented by adding a new symbol to a previously empty stack. The technical details can be found in the proof of the following theorem.

Theorem 6.10 For every Turing machine A_1 there is a two-pushdown automaton A_2 such that $L(A_2) = L(A_1)$.

PROOF First, we observe that Definition 6.19 can be modified such that all halting configurations should be of the form Uq. Namely, for any Turing machine $A_1 = (Z, K, T, M, q_0, B)$ we can introduce a new state $q_h \notin K$ and modify the mapping M such that

$$(q_h, x_0, R) \in M'(q, x) \text{ iff } M(q, x) = \varnothing$$

where x_0 is an arbitrary but fixed symbol in Z.
    Further let

$$(q_h, x, R) \in M'(q_h, x) \text{ for all } x \in Z$$

and let $M'(q_h, B) = \varnothing$. For $M(q, x) \neq \varnothing$ let $M'(q, x) = M(q, x)$.
    Having this modified Turing machine

$$A' = (Z, K \cup \{q_h\}, T, M', q_0, B)$$

we define the two-pushdown automaton

$$A'' = (Z \cup \{z_0'', B\}, K'', T \cup \{B\}, M'', z_0'', q_0, \{q_t\})$$

where

$$K'' = K \cup (K \times \{R, L\}) \cup \{q_h, q_t\}$$

and the mapping M'' is defined as follows:

1) $xqy \rightarrow xzp \in M''$ for all $x \in Z \cup \{z_0''\}$ iff $(p, z, R) \in M'(q, y)$ and $y \neq B$
2) $xqB \rightarrow x[p, R]zB \in M''$ for all $x \in Z \cup \{z_0''\}$ iff $(p, z, R) \in M'(q, B)$
3) $xqy \rightarrow x[p, L]z \in M''$ for all $x \in Z \cup \{z_0''\}$ iff $(p, z, L) \in M'(q, y)$
4) $x[p, R]z \rightarrow xzp \in M''$ for all $x \in Z \cup \{z_0''\}$, $z \in Z$, and $p \in K$
5) $x[p, L]z \rightarrow pxz \in M''$ for all $x, z \in Z$, and $p \in K$
6) $z_0''[p, L]z \rightarrow z_0''[p, L]Bz \in M''$ for all $z \in Z$, $p \in K$
7) $z_0''[q, L]B \rightarrow z_0''zp \in M''$ iff $(p, z, R) \in M'(q, B)$
8) $z_0''[q, L]B \rightarrow z_0''[p, L]z \in M''$ iff $(p, z, L) \in M'(q, B)$
9) $xq_h B \rightarrow xBq_t \in M''$ for all $x \in Z \cup \{z_0''\}$

Now let $P \in L(A')$, that is, $q_0 P \overset{*}{\underset{A'}{\Rightarrow}} Uq_h$ for some $U \in Z^*$. (Note that a halting configuration in A' always ends with q_h.) Then it is easy to see that

$$z_0'' q_0 PB \overset{*}{\underset{A''}{\Rightarrow}} z_0'' Uq_h B \underset{A''}{\Rightarrow} z_0'' UBq_t$$

and thus, $PB \in L(A'')$. (The blank symbol B is used in A'' as an end marker at the end of the word. It can be pushed further to the right if necessary.)

Conversely, if $Q \in L(A'')$ then Q must be of the form PB with $P \in Z^*$, since A'' cannot add a new B to the end of a word in Z^* and the accepting state q_t can be reached only by the move given in 9. If, on the other hand, P contained any B's then $z_0'' q_0 PB$ could never be reduced to a configuration of the form Wq_t since $M''(x, q_t, y) = \varnothing$ for all $x, y \in Z \cup \{z_0'', B\}$. Therefore, $z_0'' q_0 Q \overset{*}{\Rightarrow} z_0'' Uq_t$ implies that $Q = PB$ where $q_0 P \overset{*}{\underset{A'}{\Rightarrow}} Uq_h$.

This means that $L(A'') = L(A')\{B\}$. But then we have some type 0 grammar

$$G = (V_N, T \cup \{B\}, S, F)$$

with $L(G) = L(A')\{B\}$, and thus the grammar

$$G_B = (V_N \cup \{B\}, T, S, F \cup \{B \rightarrow \lambda\})$$

generates the language $L(A')$. Hence, by Theorem 6.8 we have some two-pushdown automaton A_2 such that $L(A_2) = L(A') = L(A_1)$ and this completes the proof.

The converse of Theorem 6.10 is also true, namely, the operation of a two-pushdown automaton can also be simulated by an appropriate Turing machine. We do not want to go into the details of this simulation, we only note that type N moves, in which the read-write head is stable, could have been

AUTOMATA AND THEIR LANGUAGES **93**

permitted for Turing machines as well. But they are not indispensable since the same effect can be achieved by introducing a new state for each type N move and make a right-shift first and then a corresponding left-shift. This would require the extension of the word by one extra square to the right where we can put an endmarker to replace the blank symbol. Erasing a symbol requires the copying of a part of the contents of the Turing tape one square to the right (or left), which is not too difficult either. Thus, we can say that Turing machines and two-pushdown automata accept the same class of languages, that is, $\mathcal{L}_0$.

EXERCISES

6.16 Develop a two-pushdown automaton that accepts the language of Exercise 1.5.

6.17 Construct a Turing machine that would write an exact copy of the input word to its right (or left) separated from it by a # symbol. (You may assume that the symbol # does not belong to $T \cup \{B\}$.) *Hint*: In order to keep track of the input symbol to be copied next we can have a position marker that can be put on top of any input symbol. In other words, we can have a secondary alphabet T' such that for each input symbol $t \in T$ the corresponding $t' \in T'$ denotes its "marked off" version. The internal states should be also used to remember the symbol to be printed next.

6.18 Construct a Turing machine that would shift the input word to right (or left) by one square on its tape. (Note that a new "quasi-blank" symbol is needed if we do not allow for printing the real blank symbol.)

6.19 Construct a Turing machine that will increment the value of its input word by one. The input and the result are to be given in binary notation on the tape.

6.20 Construct a Turing machine to compare two words separated by # on its input tape and decide whether they are identical.

SEVEN

DECIDABILITY

7.1 RECURSIVE AND RECURSIVELY ENUMERABLE LANGUAGES

A typical problem in formal language theory is to decide for some word P whether or not it belongs to a given language L. This is the so-called membership problem which is always decidable for type 1 languages as we have seen in Chapter 4. We have described a general method for deciding whether or not $P \in L(G)$ for an arbitrary word P and type 1 grammar G. The generality of the method is due to the fact that it makes no use of any specific property of P or G except for the length-increasing property of the rules of G. This means that both P and G can be considered as variables with respect to the method which can thus be applied to every P and G.

A mathematical method is said to be a *procedure* if after giving the values of the variables it can be performed fully mechanically without any risk of encountering unexpected situations. Thus, a mathematical procedure is an exact method whose details are all precisely defined in advance so that no more thinking should be required in the course of its application to any particular case. Therefore, the execution of a procedure may involve only previously defined operations and these operations should be performed in a well-defined order. In fact, apart from the limitations in time and space, every procedure can be programmed in a digital computer and conversely, every program which would run on a computer is a procedure in the above sense.

This informal concept of the mathematical procedure is not precise enough to build a meaningful mathematical theory on and later we shall see how it can be made more precise. But, interestingly enough, the above stipulated rather vague concept is still sufficient for proving a number of important properties of procedures. So we shall exploit this informal notion before giving a strictly formal definition for it which should appear more natural after these preliminary considerations.

A procedure is called an *algorithm* if it comes to a halt after a finite number of steps no matter what values are given for the variables. This is clearly a very strong requirement as there are lots of procedures which do not necessarily halt. One can design, for instance, a straightforward numerical procedure for computing the square root of a natural number with the highest possible precision. This procedure would never halt unless the integer happens to be a full square. Nevertheless, it can also be programmed for a regular computer. Such programs may run for any length of time without letting us know whether they would eventually halt. It is, therefore, very interesting to know for sure that a given procedure must always halt. This is the case, for example, with every reasonable procedure for testing the primality of a natural number.

A procedure is called a *decision procedure* if it gives only "yes" or "no" answers. A *decision algorithm* is, therefore, a procedure which always gives a "yes" or "no" answer within a finite amount of time.

A problem is said to be *algorithmically decidable* if it requires only "yes" or "no" answers and there exists a decision algorithm for it. The membership problem for context-sensitive languages, that is, the question whether or not P is in $L(G)$ is according to Theorem 4.2 algorithmically decidable. The method we presented there can be implemented as a procedure that would always halt after performing a finite number of steps. This implies, of course, that the membership problem for type 2 languages is also decidable, but for that specific case we have better algorithms as well. Usually, a problem can be solved much easier for a special case than for the general case.

For type 0 languages, however, we have left open the decidability of the membership problem. It is clear that the method we have used for context-sensitive grammars would not work for unrestricted phrase-structure grammars. Of course, this fact by itself does not imply that there is no other algorithm for it. But, unfortunately, this is truly the case with the membership problem of type 0 languages. Yet, it is far from being easy to prove such a negative statement.

As long as we are dealing with concrete algorithms only, we can be satisfied with the informal notion of algorithm. Namely, for a given procedure we can easily decide whether it is a procedure, indeed, and then we may be able to prove that it always halts. But, what can we say about arbitrary algorithms? How can we characterize them and prove theorems about all possible algorithms? Surely, we must not restrict ourselves to already established algorithms but we have to consider also the ones which would be invented only in the future. How can we specify the very concept of an algorithm without being far too restrictive? Precisely these questions have led to the development of the theory of algorithms which has become a very important branch of modern mathematics and has its impact on formal language theory as well.

So the question whether we can find an algorithm for solving the membership problem for type 0 languages can be studied only with the aid of the theory of algorithms. It is quite conceivable that this problem is really undecidable. Most likely, we expect too much, when we want to have a procedure that always decides it in a finite number of steps. But this cannot be doubtlessly concluded from the intuitive notion of an algorithm, since the latter does not tell us too much about the capability of an algorithm in general. There are some conclusions which can be drawn already from the informal notion of the algorithm and we shall do so before giving a more formal definition.

First of all, it is obvious that the decidability of the membership problem for some language L implies its decidability also for the complement of L. That is, for an arbitrary word P the relation $P \in \bar{L}$ holds if and only if $P \notin L$. Thus, we can use the same algorithm for L and $\bar{L}$; only the answers "yes" and "no" should be interpreted conversely for $\bar{L}$.

Definition 7.1 A language L is called recursively decidable or simply *recursive* iff there is an algorithm for deciding whether or not an arbitrary word belongs to L.

Definition 7.2 A language L is called *recursively enumerable* iff there is a procedure for enumerating, i.e., listing all words (possibly with repetitions) belonging to L.

The term "recursive" comes from the theory of recursive functions. It should be clear that a recursive language is also recursively enumerable. All we have to do is to generate the words of V^* one by one and each time we get a new word we use the algorithm for deciding whether or not to include that word in the enumeration. This way we get an enumerating procedure for L which, of course, would never halt for an infinite language.

On the other hand, a recursively enumerable language is not necessarily recursive. This is not quite obvious but we can show the following.

Theorem 7.1 A language L is recursive iff both L and $\bar{L}$ are recursively enumerable.

PROOF The only if part of the theorem has already been established above. To show the if part assume that both L and $\bar{L}$ are recursively enumerable.

Then combine the two enumerating procedures in such a way as to get one word alternately from each. Thus, the words of L will occur in the odd positions in this combined enumeration while the words of $\bar{L}$ will occur in the even positions.

But, since $L \cup \bar{L} = V^*$, every word $P \in V^*$ must occur somewhere in that enumeration so it can be decided in a finite number of steps whether or not $P \in L$, although this should be observed whether it occurs in an odd or even position, and this completes the proof.

If L alone is recursively enumerable and $P \notin L$, then we could wait forever to see if P appears in the enumeration. Now, the question is does there exist a recursively enumerable language L whose complement $\bar{L}$ is not recursively enumerable.

Theorem 7.2 There is a recursively enumerable language whose complement is not recursively enumerable.

PROOF Take simply a one letter alphabet $V_1 = \langle a \rangle$. Assume further that the enumerating procedures for the recursively enumerable languages over V_1 can all be described as words over a finite alphabet V_2. So we can arrange these procedures in some sequence, say,

$$P_0, P_1, \ldots, P_n, \ldots$$

where each $P_i \in V_2^*$ represents the enumerating procedure for $L_i \subseteq V_1^*$. Then let us define the language L this way

$$a^i \in L \quad \text{iff} \quad a^i \in L_i$$

It is easy to see that $\bar{L}$ cannot be recursively enumerable. For, if it were, then $\bar{L} = L_i$ should hold for some i. But that is impossible since $a^i \in L_i$ implies that $a^i \notin \bar{L}$ whereas $a^i \notin L_i$ implies that $a^i \in \bar{L}$.

It remains to show that L is recursively enumerable. First we note that every procedure performs a sequence of discrete steps which can also be numbered by the natural numbers. So we can denote by (i, j) the j-th step of the i-th procedure. If the i-th procedure comes to halt in the k-th step then the j-th step of the i-th procedure for $j > k$ will be interpreted as do nothing. Combine the procedures $P_0, P_1, \ldots$ in such a way that their steps are performed in the following order

$$(0,0), (1,0), (0,1) \; (2,0), (1,1), (0,2), (3,0), \ldots$$

Now the word a^i will be included in L iff there is a step (i, j) for some j in this sequence which includes a^i in the enumeration of L_i. Thus, we have a procedure for enumerating L which completes the proof.

Consider now the relationship of the Chomsky hierarchy to the recursive and recursively enumerable languages. We know already that every type 1 language is recursive. It should be also clear that every type 0 language is recursively enumerable. Namely, we can enumerate all words in a type 0 language simply by generating all words that are derivable from S in 1 step, 2 steps, ... Hence, we get the following theorem.

Theorem 7.3 Every type 1 language is recursive and every type 0 language is recursively enumerable.

It can be shown further that the class of type 1 languages is properly contained in the class of recursive languages.

Theorem 7.4 There exists a recursive language which is not context-sensitive.

PROOF Every type 1 grammar can be described as the list of its rules. We can assume that terminal letters do not occur on the left-hand sides of the rules, and that every nonterminal letter occurs on the left-hand side of some rule. This way the set of nonterminals can be defined implicitly by the set of rules and the initial symbol is assumed to be the left-hand side of the first rule. The set of terminals is also defined implicitly by the set of rules. Hence a word of the form

$$P_1 \to Q_1 \# P_2 \to Q_2 \cdots \# P_k \to Q_k$$

with $\#, \to \notin V_N \cup V_T$ defines a unique grammar. Consider now only those type 1 grammars which have $V_T = \{0, 1\}$ as their terminal alphabet. The choice of nonterminal symbols is immaterial for the generated language so with each grammar we can encode the nonterminals by the words

$$01, 011, 0111, \text{etc.}$$

where 01 always represents the initial symbol of the grammar. The terminal letters 0 and 1 will then be encoded by 00 and 001 while the symbols $\to$ and $\#$ will be represented by 0011 and 00111. Hence, each of our grammars will be represented by a word in $\{0, 1\}^*$. Now, all words in $\{0, 1\}^*$ can be recursively enumerated using some obvious procedure for generating those words of length 0, length 1, length 2, etc. This way we get an enumeration for our grammars as well, since we can check for each word $W_i \in \{0, 1\}^*$ whether or not it represents a type 1 grammar. So we get an enumeration of our grammars

$$G_1, G_2, \ldots, G_i, \ldots$$

which is, in fact, a subsequence of the enumeration

$$W_1, W_2, \ldots, W_i, \ldots$$

Let us define the language

$$L = \{W_i \mid W_i \notin L(G_i)\}$$

This language is clearly recursive since for each $P \in \{0, 1\}^*$ we can find within a finite amount of time the index number i for which $W_i = P$. Then we can find again in a finite number of steps the i-th grammar G_i in the sequence of grammars and finally we can decide whether or not $W_i \in L(G_i)$ since G_i is context-sensitive.

But L is not context-sensitive. If it were then it would be generated by some G_i which leads to the contradiction that $W_i \notin L(G_i)$ implies $W_i \in L(G_i)$ and vice versa. So the language L cannot be type 1 which completes the proof.

Denoting by $\mathfrak{R}$ and $\mathfrak{E}$ the classes of recursive and recursively enumerable languages, respectively, our results so far can be summarized in

$$\mathfrak{L}_1 \subset \mathfrak{R} \subset \mathfrak{E}$$

where each inclusion is proper. Further, we know that

$$\mathfrak{L}_0 \subseteq \mathfrak{E}$$

but we do not know whether this last inclusion is proper. If $\mathfrak{L}_0$ turned out to be equal to $\mathfrak{E}$ then the proper inclusion $\mathfrak{L}_1 \subset \mathfrak{L}_0$ and the undecidability of the membership problem for $\mathfrak{L}_0$ would follow immediately.

7.2 THE CHURCH-TURING THESIS

In the preceding section we have reached the question whether $\mathfrak{E} \subseteq \mathfrak{L}_0$, that is, whether all recursively enumerable languages can be generated by phrase-structure grammars. A positive answer to this question would yield the very nice equality $\mathfrak{L}_0 = \mathfrak{E}$. So we would like to show that for an arbitrary enumerating procedure there exists a type 0 grammar that accepts the enumerated language. But, what do we know about an arbitrary enumerating procedure? In the preceding proofs we have already exploited two essential properties of the informal notion of a procedure namely:

1) The action of a procedure consists of a sequence of discrete steps.
2) Every procedure has a finite description.

Let us try to formalize these two properties. First we have to specify what a step of a procedure can be like. Assume that all our data and intermediary results are written on a piece of paper as a sequence of arbitrary symbols. In each step of a procedure we may change that sequence by adding new symbols to it, erasing some symbols, or replacing them by new ones. We can assume, further, that for each step we use only a part of the data already written on the paper. Moreover, we can even assume that the procedure is broken up into such elementary steps that each would change only a single symbol on the paper by making use of a single symbol already there. Now it is clear that we have arrived at the concept of a Turing machine. Certainly, a Turing machine can be regarded as a procedure while the above reasoning is meant to support the idea that conversely, every procedure can be described in the form of a Turing machine.

This last statement is called Church-Turing thesis and it makes very deep sense. It is not a mathematical theorem for it cannot be proved by mathematical methods. The intuitive notion of the procedure is not a mathematical object so it cannot be dealt with by mathematical tools. But as soon as we identify this notion with a mathematical object like Turing machine or any other formal concept we cannot prove any more whether the informal idea has been captured correctly by the formal concept. It is possible only to disprove the Church-Turing thesis by exhibiting a procedure that is clearly a procedure in the intuitive sense but cannot be described in the form of a Turing machine. As yet, nobody has been able to construct such a counterexample. On the contrary, every other reasonable formalization of the notion of procedure such as recursive functions or Markov algorithms turned out to be equivalent to the Turing machine. This makes it reasonable to accept and use the Church-Turing thesis as a valid theorem.

In Chapter 6 we have used Turing machines as language recognizing devices which give only "yes" or "no" answers to the question whether P is in $L(A)$? Thus, we have considered them so far as decision procedures. But it is easy to define a more general procedure for every Turing machine. The contents of the tape in the initial configuration can be considered as the value of the variable(s) and the result can also be represented by the contents of the tape (or part of it) in the halting configuration. In the theory of algorithms it is usually shown how to implement the basic mathematical operations via Turing machines and how to combine two or more Turing machines into one to perform more and more complex computations. Here, we do not want to go into those details as we are concerned only with language theoretical problems.

It is, however, important to note that nondeterministic Turing machines are equivalent to deterministic ones. This can be shown by simulating the behavior of a nondeterministic Turing machine by a deterministic one. The details of this simulation will be discussed in Chapter 8 as they are related to the questions of the complexity of computations.

Returning to the class of recursively enumerable languages the next theorem will already be shown with the aid of the Church-Turing thesis.

Theorem 7.5 The class of recursively enumerable languages coincides with the class of type 0 languages.

PROOF On the basis of the Church-Turing thesis to every recursively enumerable language L there is a Turing machine that enumerates L. Without loss of generality it can be assumed that the enumerating process starts with the initial configuration $q_0 B$ and produces the sequence of configurations

$$W_1 q_f P_1, W_2 q_f P_2, \ldots, W_n q_f P_n, \ldots$$

as the listing of the words $P_i \in L$. The state q_f is a distinguished state whose appearance relates the fact that the machine has just finished with the construction of a new word. The words W_i represent the work area of the tape and they store the intermediary information for the enumerating procedure. For the above configurations we have

$$q_0 B \overset{*}{\underset{A}{\Rightarrow}} W_i p_f P_i \qquad (i = 1, 2, \ldots)$$

according to the definition of the reduction performed by the Turing machine $A = (Z, K, T, M, q_0, B)$. Let us define the phrase structure grammar $G = (V_N, V_T, S, F)$ such that

$$V_N = K \cup \{\#, \hat{q}, B, S\} \quad \text{with} \quad \#, \hat{q} \notin K \cup Z, V_T = Z$$

and the rules in F are the following:

1) $S \to \# q_0 B \in F$
2) $P \to Q \in F$ if $P \to Q \in M$
3) $q_f \to \hat{q} \in F$
4) $z\hat{q} \to \hat{q} \in F$ for all $z \in Z$
5) $\#\hat{q} \to \lambda \in F$

It is easy to see that $S \overset{*}{\underset{G}{\Rightarrow}} P$ holds for some $P \in T^*$ iff $q_0 B \overset{*}{\underset{A}{\Rightarrow}} W q_f P$ holds for some W. Therefore, the class $\mathcal{E}$ is contained in $\mathcal{L}_0$, which combined with $\mathcal{L}_0 \subseteq \mathcal{E}$ gives the result.

Note that the Turing machine A in the above proof can be assumed to be deterministic.

In Section 7.1 we have seen that there is a recursively enumerable language whose complement is not recursively enumerable. In that proof we did not use the Church-Turing thesis. Now, using the Church-Turing thesis it can be translated into a theorem about Turing machines. It says, namely, that there exists a type 0 language whose complement is not type 0, that is, there exists a Turing machine that accepts a language whose complement cannot be accepted by any Turing machine. In this form the theorem can be shown again independently of the Church-Turing thesis. The proof is essentially the same as that of Theorem 7.2 only we have to diagonalize over the set of all Turing machines which is denumerable so it can be arranged into a sequence

$$T_1, T_2, \ldots, T_k, \ldots$$

For the sake of simplicity we can take again a one letter alphabet $V = \{a\}$ and define the language L such that $a^k \in L$ if $a^k \in L(T_k)$. Then the language $\bar{L}$ cannot be accepted by any Turing machine.

This, of course, shows only that the Church-Turing thesis is consistent with the rest of our theory, which is no wonder. Thus, we can first prove the theorem for Turing machines and then use the Church-Turing thesis to conclude that there exists a type 0 language whose complement is not recursively enumerable in the informal sense. By Theorem 7.1 this means that this type 0 language cannot be recursive. In other words, the membership problem for arbitrary type 0 languages is undecidable. Now, as the relation $P \in L(T)$ is equivalent to the condition that for the input P the machine T comes to a halt after a finite number of steps, the above theorem is often expressed in the form that the halting problem for Turing machines is undecidable.

According to the Church-Turing thesis this can be converted again into a theorem about Turing machines as follows.

Theorem 7.6 There is no Turing machine that could decide for arbitrary word P and Turing machine T whether or not T halts for P.

In this form the theorem is obviously independent of the Church-Turing thesis so it can be shown without it. Because of the importance of this theorem we also present a standard proof for it although, as we have said before, it is a consequence of Theorem 7.5 and the Church-Turing thesis.

PROOF Every Turing machine T can be described as a finite word. This word can be encoded as T' in the tape alphabet of T. A Turing machine is called self-applicable if when started with the initial configuration $q_0 T'$ it comes to a halt after a finite number of steps. Let us make the indirect assumption that a Turing machine U exists for deciding the halting problem for every Turing machine. Then U could decide the self-applicability problem, as well, which is but a special case of the halting problem. In order for U to be able to decide the self-applicability of an arbitrary Turing machine T the latter should be encoded as T'' in the tape alphabet of U. Now, let the initial state of U denoted by q_0 and let q_1 and q_2 be two distinguished states of U such that

$$q_0 T'' \underset{U}{\overset{*}{\Rightarrow}} X_1 q_1 Y_1 \quad \text{iff } T \text{ is self-applicable}$$

$$q_0 T'' \underset{U}{\overset{*}{\Rightarrow}} X_2 q_2 Y_2 \quad \text{iff } T \text{ is not self-applicable}$$

where $X_1 q_1 Y_1$ and $X_2 q_2 Y_2$ are some halting configurations of U. Construct a Turing machine $\bar{U}$ that halts for the input T'' if and only if T is not self-applicable. This can be done by modifying U in such a way that a new state q_3 is introduced together with the moves

$$q_1 z \to z q_3 \quad \text{for all } z \text{ with } M(q_1, z) = \varnothing$$

and

$$q_3 x \rightarrow z q_3 \quad \text{for all } x \in Z \cup \{B\}$$

where z is an arbitrary but fixed symbol of Z.

What happens now if $\overline{U}$ is applied to itself. It should never halt when it is self-applicable and it should halt when it is not which contradicts the definition of self-applicability. This completes the proof.

7.3 UNDECIDABLE PROBLEMS

The undecidability of the membership problem for type 0 languages has far reaching consequences. A large number of problems can be shown to be undecidable by reducing them to the membership problem of type 0 languages or, equivalently, to the halting problem of Turing machines. In fact, even for the context-sensitive or context-free case, most of the interesting questions about languages are undecidable. We shall give here only a few examples.

Theorem 7.7 It is undecidable whether an arbitrary type 1 language is empty.

PROOF Let $G = (V_N, V_T, S, F)$ be an arbitrary type 0 grammar and P be an arbitrary word in V_T^*. Construct the type 1 grammar

$$G_1 = \left(V_N \cup V_T \cup \{S_0, S_1, S_2\}, \{\#\}, S_0, F_1 \right)$$

where S_0, S_1, S_2, and $\#$ are not in $V_N \cup V_T$ and the rules in F_1 are the following:

1) $S_0 \rightarrow S_1 S S_2 \in F_1$
2) $X \rightarrow Y \in F_1$ if $X \rightarrow Y \in F$ and $|X| \leqslant |Y|$
3) $X \rightarrow Y \#^k \in F_1$ if $X \rightarrow Y \in F$ and $|X| = |Y| + k$ with $k > 0$
4) $\#u \rightarrow u\# \in F_1$ for all $u \in V_N \cup V_T$
5) $S_1 P \rightarrow \#^m S_2 \in F_1$ where $m = |P|$ for the P in question
6) $S_2\# \rightarrow \#S_2 \in F_1$ and $S_2 S_2 \rightarrow \#\# \in F_1$

It is easy to see that $P \in L(G)$ iff $L(G_1)$ is not empty. Thus, any algorithm that would decide the emptiness of an arbitrary type 1 language could also be used for deciding the membership problem for type 0 languages.

Theorem 7.8 It is undecidable whether an arbitrary type 1 language is infinite.

PROOF Let $G = (V_N, V_T, S, F)$ be again an arbitrary type 0 grammar and

$P \in V_T^*$. Construct the type 1 grammar

$$G_1 = (V_N \cup V_T \cup \{S_0, S_1, S_2\}, \{\#\}, S_0, F_1 \cup \{S_2 \rightarrow S_2\#\})$$

where $S_0, S_1, S_2, \# \notin V_N \cup V_T$, and the rules in F_1 are defined in the same way as in Theorem 7.7. Now $P \in L(G)$ iff the language $L(G_1)$ is infinite.

Theorem 7.9 It is undecidable whether the intersection $L_1 \cap L_2$ of two arbitrary type 2 languages is empty.

PROOF As we have seen, the intersection of two type 2 languages is not always type 2. Now, we shall prove that an arbitrary type 0 language can be expressed as the homomorphic image of the intersection of two type 2 languages.

Let $G = (V_N, V_T, S, F)$ be an arbitrary type 0 grammar and let the rules in F be supplied with labels $f_i \in V_F$, in the same way as given in Section 5.2. Thus, each rule in F will have a unique label, in symbols

$$f_i : P_i \rightarrow Q_i \in F, \qquad 1 \leqslant i \leqslant n$$

where the set of labels is

$$V_F = \{f_1, f_2, \ldots, f_n\}$$

and $V_F \cap (V_N \cup V_T) = \varnothing$.

Next we define the alphabet V_T' such that $V_T' \cap (V_F \cup V_N \cup V_T) = \varnothing$ and there is a one-to-one correspondence between V_T and V_T', in symbols

$$a' \in V_T' \quad \text{iff} \quad a \in V_T$$

Now, we define two languages

$$L_1 = \{ Xf_i P_i Y \# Y^{-1} Q_i^{-1} X^{-1} \# \mid X, Y \in (V_N \cup V_T)^*, 1 \leqslant i \leqslant n \}^+ (V_T')^*$$

$$L_2 = \{ f_i S \# \mid 1 \leqslant i \leqslant n \} \{ Y^{-1} X^{-1} \# Xf_i Y \# \mid X, Y$$
$$\in (V_N \cup V_T)^*, 1 \leqslant i \leqslant n \}^* \{ P^{-1} \# P' \mid P \in V_T^* \},$$

where $\# \notin V_T' \cup V_F \cup V_N \cup V_T$ and P' denotes the word in $(V_T')^*$ that corresponds to $P \in V_T^*$. Each of these languages is context-free which follows from the fact that the languages

$$\{ Xf_i P_i Y \# Y^{-1} Q_i^{-1} X^{-1} \# \mid X, Y \in (V_N \cup V_T)^* \} \qquad (i = 1, \ldots, n)$$

and

$$\{ Y^{-1} X^{-1} \# Xf_i Y \# \mid X, Y \in (V_N \cup V_T)^* \} \qquad (i = 1, \ldots, n)$$

as well as

$$\{ P^{-1} \# P' \mid P \in V_T^* \}$$

are all context-free and $\mathcal{L}_2$ is closed under the regular operations. (For instance, the rules $S \rightarrow A\#$, $A \rightarrow f_i P_i B Q_i^{-1}$, $B \rightarrow \#$, and for all x in $V_N \cup V_T$, $A \rightarrow xAx$ and $B \rightarrow xBx$ can be used to generate the first of these languages.)

The words in $L_1 \cup L_2$ will represent the derivations in G since they have the forms

$$f_{i_1}S\#Q_{i_1}^{-1}\#X_2 f_{i_2}P_{i_2}Y_2\#Y_2^{-1}Q_{i_2}^{-1}X_2^{-1}\#X_3 f_{i_3}P_{i_3}Y_3\# \cdots \#P^{-1}\#P'$$

where

$$X_{k+1}P_{i_{k+1}}Y_{k+1} = X_k Q_{i_k}Y_k \qquad \text{for } k = 1, 2, \ldots$$

which is equivalent to the existence of a derivation

$$S \underset{G}{\overset{*}{\Rightarrow}} P$$

for the given P.

Let us define the homomorphism h such that

$$h(a') = a \in V_T \quad \text{for all } a' \in V_T'$$
$$h(x) = \lambda \quad \text{for all } x \in V_F \cup V_N \cup V_T \cup \{\#\}$$

Hence, the words in $h(L_1 \cap L_2)$ will be exactly those in $L(G)$ and thus, $L(G)$ is empty iff $L_1 \cap L_2 = \varnothing$ which completes the proof.

The proof of the above theorem is based on the fact that the set of derivations in an arbitrary type 0 grammar can be represented by the intersection of two context-free languages. (In Chapter 10 we shall see that it can also be represented by a deterministic context-sensitive language.) If we use Turing machines in place of type 0 grammars we obtain the result that the set of reductions (computations) of a Turing machine can be represented by the intersection of two context-free languages. As we know, the intersection of context-free languages is not context-free in general. The complement of the above intersection is, however, context-free as shown below.

Theorem 7.10 For any Turing machine the set of invalid reductions (computations) can be represented by a context-free language.

PROOF Let $A(Z, K, T, M, q_0, B)$ be an arbitrary Turing machine and let $\#$ be an extra symbol not in $Z \cup K \cup T \cup \{B\}$. A valid reduction can be represented by a word satisfying the following two conditions:

1) It has the form
$$W_0\#W_1^{-1}\#W_2\# \cdots \#W_n^{(-1)^n}$$
where each W_i is a configuration of A and W_0 is the initial configuration while W_n is a halting configuration.
2) $W_i \underset{A}{\Rightarrow} W_{i+1}$ holds for $i = 0, 1, \ldots, n - 1$.

Hence, an invalid reduction can be represented by a word which does not satisfy at least one of these two conditions.

It is easy to construct a finite automaton to check for condition 1. Thus, the set of words which do not satisfy condition 1 is also a type 3 language.

To check for the negation of condition 2 we need a nondeterministic pushdown automaton which would skip everything on the input tape until it nondeterministically selects an occurrence of #. Then it starts reading the word

$$W_i^{(-1)^i}$$

and simultaneously pushing a word $U^{(-1)^i}$ onto the stack for which

$$W_i \underset{A}{\Rightarrow} U$$

holds. (Note that the parity of i can be remembered by the finite state control of the pushdown store machine.) Thereafter it compares its stack with

$$W_{i+1}^{(-1)^{i+1}}$$

and if they match then it rejects the input, otherwise it accepts the input.

Hence, the set of invalid reductions is represented by the union of a regular and a context-free language which is context-free.

Corollary 1 For an arbitrary alphabet V and for an arbitrary context-free grammar G it is undecidable whether or not $V^* = L(G)$.

PROOF For an arbitrary Turing machine A we can find a context-free grammar G and an alphabet V such that $L(A) = \varnothing$ iff $V^* = L(G)$. Namely, $L(A)$ is empty iff the set of valid reductions is empty, i.e., the set of invalid reductions is V^* for the appropriate alphabet V. Hence, if $V^* = L(G)$ were always decidable then the emptiness problem of the type 0 languages would be decidable as well, contrary to Theorem 7.7.

Corollary 2 For arbitrary context-free grammars G_1 and G_2 and for an arbitrary regular language R the following problems are undecidable.

1) Is $L(G_1) = L(G_2)$?
2) Is $L(G_1) \subseteq L(G_2)$?
3) Is $R = L(G_2)$?
4) Is $R \subseteq L(G_2)$?

PROOF Clearly, problem 3 is a special case of 1, and 4 is a special case of 2. Further, $V^* = L(G_2)$ is a special case of both 3 and 4. Hence, none of these problems is decidable.

As we have mentioned earlier, there are many other interesting problems which are undecidable. Some of the proofs of these unsolvability results are fairly complex. Here, we do not want to present results without proofs, therefore, the interested reader is referred to the literature. In short, we may say that almost every interesting problem is unsolvable for type i languages except for type 3.

Given the fact that so many problems are unsolvable, one is tempted to try to increase the capability of the Turing machine by using more than one tape or two-dimensional tapes, or more read-write heads, etc. Unfortunately, it does not help; the unsolvable problems remain the same as with the simplest model. This is again an indirect evidence for the validity of the Church-Turing thesis.

The only way in which the power of the Turing machine can be increased is to allow for infinitely many steps. The languages whose membership problems can be decided in denumerably infinite steps are precisely the recursively enumerable languages. An interesting development of the theory is concerned with the convergence of the infinite computations and with the continuity of the computable functions (see, e.g., Stoy [1977]).

In the following chapter we shall see that the solution of some theoretically decidable problems may require such a tremendous amount of time (and/or storage space) that solution is practically unacceptable. Therefore, the class of decidable problems should be even further restricted if we do not allow for an arbitrarily large number of steps. So we would ask questions like what problems are decidable within a limited amount of time? The time limit will usually be given as a function of the problem itself rather than a constant.

EXERCISES

7.1 Design a Turing machine that enumerates the language

$$\{ PP \mid P \in \{a, b\}^* \}$$

7.2 Prove that $\mathcal{L}_0$ is closed under intersection.

7.3 Prove that the class of recursive languages $\mathcal{R}$ is closed under intersection, union, and complementation.

7.4 Prove that $\mathcal{L}_1$ is closed under intersection.

EIGHT

COMPLEXITY OF COMPUTATIONS

8.1 DETERMINISTIC AND NONDETERMINISTIC PROCEDURES

It seems to be useful to prove that the computing power of deterministic procedures is not less than that of the nondeterministic ones. In other words, it would be important to show that whatever is computed by a nondeterministic procedure can be computed by a deterministic procedure as well. A mathematical proof to this effect can be given, of course, only for some formal model such as the Turing machine. But, before presenting any proof, it may be worthwhile to see again what reason we have for dealing with nondeterministic procedures at all. Deterministic procedures appear to be much simpler in general, but it should be borne in mind that only equivalent procedures should be compared, i.e., those that would do the same job.

A sequence of steps performed by a procedure when applied to a given input is called a *computation*. If the steps of a computation are represented as the edges of a directed graph, then a deterministic computation can be visualized as a chain of such edges, whereas a nondeterministic computation corresponds to a directed tree in which the edges starting from the same node represent the possible choices of the next step in the course of the computation. Therefore, the set of all possible sequences of steps in a nondeterministic computation corresponds to the set of all directed paths starting from the root of the tree.

A nondeterministic computation is said to be *terminating* if it has at least one terminating sequence of steps (i.e., a sequence of steps that leads to a halting configuration). Thus, the different possible sequences of steps in a nondeterministic computation can be regarded as various trials in a search procedure in which the solution is represented by a terminating sequence and there may be any number of possibly nonterminating sequences. In this sense nondeterministic procedures are quite natural, and in many cases they may even be simpler than their deterministic counterparts.

Take, for instance, a chess puzzle in which we are given a starting position and asked to find a sequence of moves that ensures the checkmating of the opposite king. For this problem we may set up one chessboard and successively try out all possible moves, keeping records of completed trials and returning to earlier positions when necessary. But we can also imagine that we set up as many chessboards as there are possibilities for the first move. Then for each chessboard we repeat this process, adding as many chessboards as there are possibilities for each succeeding move. If the puzzle does have a solution, then after a finite number of moves we must find a checkmating position on one of the chessboards. (In fact, the picture is a bit more complex, since we have to take into account the moves of the opponent as well.) The number of necessary chessboards may increase sharply with the number of moves, but it is easier to think of such an arbitrary number of chessboards and design a nondeterministic procedure for solving the problem than to think of one chessboard and design a deterministic procedure for solving the same problem in a strictly sequential manner. In fact, the problem as stated is essentially nondeterministic since nothing has been said about the order in which to make the trials. Thus, to treat this problem in a deterministic way we have to execute the conceptually parallel moves of a nondeterministic process in some artificially well-defined order.

Similarly, there are many problems that can be formulated more simply in terms of nondeterministic Turing machines. This is the case, for example, with the proof of the equivalence of type 0 grammars and Turing machines which has been discussed in Chapter 6. Now, we shall prove that every nondeterministic Turing machine can be simulated by a deterministic one in the sense that any possible sequence of moves of the former will be executed sometime by the latter. For this purpose, we introduce first the notion of the multitape Turing machine (see Figure 8.1) which is a direct generalization of the single-tape

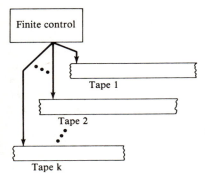

Figure **8.1** k-tape Turing machine.

model. A k-tape Turing machine has k tapes and k read-write heads, each scanning a different tape. Each move of the machine consists of changing the internal state of the finite state control and performing simultaneously for each tape such elementary operations as have been defined for the single-tape case. (These operations are, in general, different for each tape.)

Definition 8.1 A k-tape Turing machine is an ordered $(k + 4)$-tuple $A = (Z_1, \ldots, Z_k, K, M, B, q_0)$

where Z_i is the *tape alphabet* of the i-th tape $(i = 1, \ldots, k)$
K is the finite set of *internal states*
$B \notin Z_i$ $(i = 1, \ldots, k)$ is the blank symbol
$q_0 \in K$ is the *initial state*
M is a mapping from $K \times (Z_1 \cup \{B\} \times \cdots \times (Z_k \cup \{B\}))$ into the set of subsets of $K \times Z_1 \times \cdots \times Z_k \times \{L, R\}^k$, called the *transition function*

According to this definition, for $p, q \in K$, $x_i \in Z_i \cup \{B\}$, $y_i \in Z_i$, and $t_i \in \{L, R\}$ $(i = 1, \ldots, k)$ the relation

$$(p, y_1, \ldots, y_k, t_1 \cdots t_k) \in M(q, x_1, \ldots, x_k)$$

means that the nondeterministic k-tape Turing machine, when being in state q and scanning x_i on its i-th tape, is able to change its state to p together with printing y_i on tape i and shifting its read-write head as specified by the value of t_i (for $i = 1, \ldots, k$).

Theorem 8.1 Every k-tape Turing machine can be simulated by a one-tape Turing machine.

PROOF First we map the contents of the k separate tapes onto one tape. To this end, we divide the tape of the one-tape Turing machine into $2k$ (imaginary) tracks. (See Figure 8.2.) The odd-numbered tracks correspond to the tapes of the multitape machine, while each even-numbered track contains all blanks except for one occurrence of the symbol # denoting the current position of the corresponding read-write head. Without loss of generality we can assume that these position markers occupy the same position (i.e., they are in the same column) in the initial configuration of the simulating one-tape machine.

Now, given a k-tape Turing machine $A = (Z_1, \ldots, Z_k, K, M, B, q_0)$, we define the simulating one-tape machine $A' = (Z', K', M', B', q_0')$ such

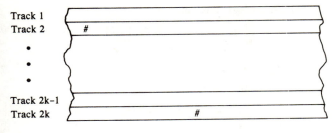

Track 1
Track 2

Track 2k-1
Track 2k

Figure 8.2

that

$$B' = [B, B, \ldots, B]$$

$$Z' = (Z_1 \cup \{B\}) \times \{B, \#\} \times \cdots \times (Z_k \cup \{B\}) \times \{B, \#\} - \{B'\}$$

$$K' = K \times (Z_1 \cup \{B, ?\}) \times \cdots \times (Z_k \cup \{B, ?\}) \times \{L, R, 0, 1\}^k$$

$$q_0' = [q_0, ?, \ldots, ?, 0 \cdots 0]$$

and the transition function M' is defined as follows. (Note that each state of A' can store a tape symbol and a direction symbol for each of the original tapes. The symbols $? \notin Z_i$ and 0 represent here the lack of such information.)

1) For all $q \in K$ and $[z_1, m_1, \ldots, z_k, m_k] \in Z'$ let

$$([q, s_1', \ldots, s_k', 0 \cdots 0], [z_1, m_1, \ldots, z_k, m_k], R)$$
$$\in M'([q, s_1, \ldots, s_k, 0 \cdots 0], [z_1, m_1, \ldots, z_k, m_k])$$

iff $s_i = ?$ for at least one value of i $(1 \leqslant i \leqslant k)$. Here the values of s_i' are defined as

$$s_i' = \begin{cases} z_i & \text{if } m_i = \# \\ s_i & \text{otherwise} \end{cases}$$

2) For all $z_1, \ldots, z_k$ with $z_i \in Z_i \cup \{B\}$ let

$$([p, y_1, \ldots, y_k, t_1 \cdots t_k], [z_1, B, \ldots, z_k, B], L)$$
$$\in M'([q, x_1, \ldots, x_k, 0 \cdots 0], [z_1, B, \ldots, z_k, B])$$
$$\text{iff } (p, y_1, \ldots, y_k, t_1 \cdots t_k) \in M(q, x_1, \ldots, x_k)$$

3) For all $p \in K$, $[z_1, m_1, \ldots, z_k, m_k] \in Z'$, $t_1 \ldots t_k \in \{L, R, 1\}^k$, and $s_1, \ldots, s_k$ with $s_i \in Z_i \cup \{B, ?\}$ such that $s_i \neq ?$, $m_i = \#$, and $t_i = R$

hold simultaneously for at least one value of i let

$$([p, s'_1, \ldots, s'_k, t'_1 \cdots t'_k], [z'_1, m'_1, \ldots, z'_k, m'_k], R)$$
$$\in M'([p, s_1, \ldots, s_k, t_1 \cdots t_k], [z_1, m_1, \ldots, z_k, m_k])$$

where
$$z'_i = \begin{cases} s_i & \text{if } s_i \neq ?, \quad m_i = \#, \quad \text{and } t_i = R \\ z_i & \text{otherwise} \end{cases}$$

$$m'_i = \begin{cases} B & \text{if } s_i \neq ?, \quad m_i = \#, \quad \text{and } t_i = R \\ \# & \text{if } s_i \neq ?, \text{ and } t_i = 1 \\ m_i & \text{otherwise} \end{cases}$$

$$s'_i = \begin{cases} ? & \text{if } s_i \neq ?, \text{ and } t_i = 1 \\ s_i & \text{otherwise} \end{cases}$$

$$t'_i = \begin{cases} 1 & \text{if } s_i \neq ?, \quad m_i = \#, \quad \text{and } t_i = R, \\ t_i & \text{otherwise} \end{cases}$$

4) For all $p \in K$, $[z_1, m_1, \ldots, z_k, m_k] \in Z'$, $t_1 \cdots t_k \in \{L, R, 1\}^k$, and $s_1, \ldots, s_k$ with $s_i \in Z_i \cup \{B, ?\}$ such that $s_i \neq ?$ for at least one value of i, but $s_i \neq ?$, $m_i = \#$, and $t_i = R$ do not hold simultaneously for any i, let

$$([p, s'_1, \ldots, s'_k, t'_1 \cdots t'_k], [z'_1, m'_1, \ldots, z'_k, m'_k], L)$$
$$\in M'([p, s_1, \ldots, s_k, t_1 \cdots t_k], [z_1, m_1, \ldots, z_k, m_k])$$

where
$$z'_i = \begin{cases} s_i & \text{if } s_i \neq ?, \quad m_i = \#, \quad \text{and } t_i = L \\ z_i & \text{otherwise} \end{cases}$$

$$m'_i = \begin{cases} B & \text{if } s_i \neq ?, \quad m_i = \#, \quad \text{and } t_i = L \\ \# & \text{if } s_i \neq ?, \text{ and } t_i = 1 \\ m_i & \text{otherwise} \end{cases}$$

$$s'_i = \begin{cases} ? & \text{if } s_i \neq ?, \text{ and } t_i = 1 \\ s_i & \text{otherwise} \end{cases}$$

$$t'_i = \begin{cases} 1 & \text{if } s_i \neq ?, \quad m_i = \#, \quad \text{and } t_i = L \\ t_i & \text{otherwise} \end{cases}$$

5) For all $p \in K$, and $z_1, \ldots, z_k$, with $z_i \in Z_i \cup \{B\}$ let

$$([p, ?, \ldots, ?, 0 \cdots 0], [z_1, B, \ldots, z_k, B], R)$$
$$\in M'([p, ?, \ldots, ?, 1 \cdots 1], [z_1, B, \ldots, z_k, B])$$

On the basis of this construction the simulation takes place as follows. Each move of the k-tape Turing machine is implemented by a series of moves called a basic cycle of the simulating one-tape Turing machine. The first part of this cycle, called phase 0, will scan the tape from left to right between the leftmost and rightmost markers in order to read the symbols occurring at the markers and store them into the s_i. Phase 0 will use only type 1 moves, i.e., those given by part 1 of the above construction. As soon as all s_i's are filled up, a type 2 move will compute the next move for the original k-tape machine and will shift back the read-write head to the position of the rightmost marker(s), which was (were) left behind by one square in the last move of phase 0. (For simplicity's sake the simulating machine is permitted to print B' in a type 2 or 5 move when all z_i are blanks.) Now, in phase 1 the moves specified by 3 and 4 will take care of printing the new symbols onto their appropriate places and shifting each marker to the right or left as it is required. When more than one track contains markers in the same position then those that have to be shifted to the right are processed first. Such markers are processed simultaneously by a type 3 move. Here $s_i' \neq ?$ and $t_i' = 1$ will ensure that a marker is printed on the corresponding track in the next move. (Immediately after a type 3 move no other type 3 move can follow, since $m_i = \#$ and $t_i = R$ cannot hold for any i at such a place. This can be shown by induction starting from the rightmost marker(s).) The type 4 move coming next will then print the markers for the tracks with $s_i \neq ?$ and $t_i = 1$ and clear the value of s_i by storing the symbol $?$ in it, which means that the corresponding track has been completed. Type 4 moves will process at the same time all tracks with $s_i \neq ?$, $m_i = \#$, and $t_i = L$. After having finished with the leftmost marker(s), a type 5 move leads back to the beginning of phase 0 and shifts the read-write head back to the leftmost marker(s).

Now we are in better shape to deal with the connection between deterministic and nondeterministic Turing machines. In the rest of this chapter we shall give only the outlines of the proofs, leaving out most of the technical details. The construction of the preceding proof makes it clear that a deterministic k-tape Turing machine can always be simulated by a deterministic one-tape Turing machine. This fact will be used in the corollary of the next theorem.

Theorem 8.2 Every nondeterministic one-tape Turing machine can be simulated by a deterministic 3-tape Turing machine.

PROOF The first tape of the deterministic Turing machine will be used to store the input word. The second tape will be used to determine all possible sequences of moves of the nondeterministic machine in some well-defined order. The third tape will be used as a work tape to simulate the computation of the nondeterministic machine for each particular sequence of choices of moves generated on tape 2.

Let the nondeterministic Turing machine have at most r possible moves for each state-input pair and let V be an arbitrary r-letter alphabet. Then a sequence of m consecutive moves can be represented by a word $P = a_1 \cdots a_m \in V^*$ where a_i represents the actual choice made for the i-th move in the sequence.

For the deterministic simulation we generate all words in V^* one by one, starting with those of length 1, then of length 2, etc. For each such word we copy the input from tape 1 onto tape 3 and simulate the operation of the nondeterministic machine with respect to the choices of moves specified by the word on tape 2. If the next letter of that word does not specify a valid choice (the current state-input pair may have less than r moves) then we return to the generation of the next word of V^* on tape 2. This way we can simulate all possible sequences of moves of the nondeterministic Turing machine. If the latter has a terminating sequence then this sequence will be executed also by the deterministic machine within a finite amount of moves.

Corollary Every nondeterministic Turing machine can be simulated by a deterministic one-tape Turing machine.

This corollary follows simply from Theorems 8.1 and 8.2. It should be observed, however, that the length of the word generated on tape 2 depends on the number of steps to be performed, and thus it need not be linearly bounded for a linear bounded automaton. Therefore, the equivalence of deterministic and nondeterministic linear bounded automata cannot be shown in this fashion. For that matter, it has not been either proved or disproved by anyone as yet and it is certainly one of the most famous open problems in the theory of automata (see also Theorems 8.3 and 8.4).

8.2 MEASURES OF COMPLEXITY

In the preceding section nondeterministic Turing machines have been reduced to deterministic ones, but this reduction has been achieved at the cost of increasing the size of the alphabet and the number of moves, etc. It would be, therefore, interesting to know what is the simplest possible way of solving some problem. First of all we should know how to compare two procedures to decide which is simpler or, equivalently, which is more complex than the other.

There are several ways of defining the complexity of procedures. The *static* or *definitional complexity* of a mathematical system can be measured by the number of its components. The definitional complexity of a Turing machine would thus be its number of moves, i.e., the cardinality of the set of the instructions making up the machine. This is the same as the length of the program in case of ordinary programming languages.

The *dynamic complexity measures*, on the other hand, are related to the running time and the storage space required by the actual computation. Obviously, these depend on the input data, too, so they should be expressed as functions of the input values. In order to avoid the complications involved in the interpretation of the particular data, a usual approach, adopted here, is to consider only the total length of the data. For Turing machines this means simply the length of the input word. The running time will be defined as the number of moves performed by the Turing machine in the course of the computation, while the storage space can be measured by the amount of the tape squares used for the given computation. So we shall use the following definitions:

A Turing machine is said to be of time-complexity $t(n)$ if it halts in at most $t(n)$ moves for all input words of length at most n.

A nondeterministic Turing machine is considered to halt in at most $t(n)$ moves if there is at least one terminating sequence of moves (i.e., leading to a halting configuration) with at most $t(n)$ moves. (If there is no terminating sequence of moves for some input word of length $\leqslant n$, then $t(n)$ is obviously infinite.)

A Turing machine is said to be of space-complexity $s(n)$ if for all input words of at most n length it has a terminating sequence of moves which makes use of at most $s(n)$ tape squares on each of its tapes.

Clearly, the functions $t(n)$ and $s(n)$ are monotonic by definition. These definitions are given for multitape Turing machines described in Section 8.1. However, for space-complexity measures it is sometimes better to use a different model (see Figure 8.3) where we have a read-only input tape in addition to the k work tapes. The reason for using this model is to allow for

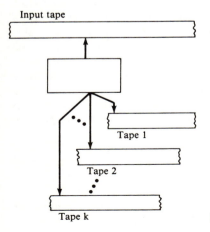

Figure 8.3

cases $s(n) < n$, so that the amount of the space used strictly for computation may be less than the space containing the input. It is reasonable to assume that the whole input word is always read, otherwise its total length should not be used as the argument of the complexity functions.

Now, if we are given two deterministic one-tape Turing machines solving the same problem, it still can happen that for some input lengths the first is simpler (i.e., takes fewer moves or uses less space) than the second while for other input lengths just the opposite is true. In order to reduce this kind of ambiguity, a finite number of exceptions will be tolerated, which means that only the asymptotical behavior of the complexity functions will be examined. Thus, for example, the time complexity of the first Turing machine is said to be greater than that of the other iff there is some natural number n_0 such that $t_1(n) > t_2(n)$ for all $n > n_0$.

The above introduced dynamic measures are the most interesting ones, and they are also relevant to the design of practical algorithms. For every algorithmically solvable problem it would be nice to know which is the best (i.e., the least complex in some sense) algorithm for solving it. Unfortunately, this is a far too optimistic idea. A somewhat more realistic approach is to classify the problems according to some well-defined functions, such as $\log n$, n, n^2, e^n, and so forth, as upper bounds on the amount of time or space that can be used for solving the problems belonging to that class. Then we can try to find the smallest complexity class to which a given problem belongs.

Here we are interested mainly in language recognition problems, so we shall define the complexity classes as language classes with certain functions as upper bounds to the complexity of their recognition problems. This language theoretical approach is also justified by the fact that a great variety of problems from different branches of mathematics can be considered as language recognition problems. Decision problems require only "yes" or "no" answers for each particular case, therefore, we can encode each instance of such a problem as a word over a fixed alphabet and reformulate our decision problem as the recognition of those words for which the answer is "yes." Take, for example, the decision of the primality of the natural numbers. By using decimal (or some other) notation this will be equivalent to the recognition of the language consisting precisely of the primes in the given notation. More general problems may also be converted into decision problems by guessing a solution first and then checking it. Of course, the guessing part of the decision procedure should be made equivalent to an exhaustive search for all possible solutions.

The class of all languages that can be recognized by deterministic k-tape Turing machines of time-complexity $t(n)$ such that $t(n) \leqslant f(n)$ for sufficiently large n will be denoted by DTIME($f(n)$).

Similarly, NTIME($f(n)$), DSPACE($f(n)$), and NSPACE($f(n)$), respectively, denote the nondeterministic time, deterministic space, and nondeterministic space-complexity classes bounded by the function $f(n)$.

Note that $f(n) \geqslant n$ must hold for both DTIME($f(n)$) and NTIME($f(n)$), otherwise the Turing machine cannot completely read the input. Further, it should be clear that DTIME($f(n)$) $\subseteq$ DSPACE($f(n)$) and NTIME($f(n)$) $\subseteq$ NSPACE($f(n)$), because a Turing machine cannot use more squares from any of its tapes than the number of moves it makes.

The above defined complexity classes are all proper subsets of the class of recursive (decidable) languages $\Re$. To put it differently, *there is no computable total function $f(n)$ such that $\Re \subset NSPACE(f(n))$*. This can be shown indirectly by using diagonalization over the set of all Turing machines. The proof is essentially the same as with Theorem 7.4, only we have to enumerate all nondeterministic Turing machines instead of all type 1 grammars and define the language

$$L = \{W_i | W_i \notin L(T_i)\}$$

This language cannot be accepted by any Turing machine and still, it would be decidable because we can always find the index of a word W in the enumeration of V^* and thus, we can also find the corresponding Turing machine in the enumeration of the latter. Finally, we can compute the value of $f(|W|)$ and check whether W is accepted by the corresponding Turing machine within $f(|W|)$ space. (The finite amount of space allows only for a finite number of different configurations which in turn gives the upper limit of the length of the sequences of moves to be checked for. Longer sequences of moves must produce repetitions of the same configurations in the course of the computation so they can be ignored when looking for a terminating sequence of moves.) Hence, we would have a decision algorithm for L which, in view of the Church-Turing thesis, contradicts the fact that L is not accepted by any Turing machine. Now, since

$$DSPACE(f(n)) \subseteq NSPACE(f(n))$$

and

$$DTIME(f(n)) \subseteq NTIME(f(n)) \subseteq NSPACE(f(n))$$

the set of all recursive languages $\Re$ is not included in any of these complexity classes.

This result implies that there are infinite hierarchies of space- and time-complexity classes corresponding to functions growing faster and faster with n. However, a more detailed study of these classes would reveal that there must be some gap between $f(n)$ and $g(n)$ in order to have a proper inclusion between the corresponding two classes. In particular, it can be shown that for any constant $c > 0$

$$DSPACE(cf(n)) = DSPACE(f(n))$$

and

$$NSPACE(cf(n)) = NSPACE(f(n))$$

and also

$$DTIME(cf(n)) = DTIME(f(n)) \quad \text{and} \quad NTIME(cf(n)) = NTIME(f(n))$$

These results are called the linear tape-compression and linear speed-up theorems, respectively.

Theorem 8.3 If L is accepted by an $S(n)$ space-bounded Turing machine with k work tapes (see Figure 8.3), then for any $c > 0$, L is accepted by a $[cS(n)] + 1$ space-bounded Turing machine also with k work tapes.*

PROOF The operation of the $S(n)$ space-bounded Turing machine can be simulated by another Turing machine where each tape square represents a block of m adjacent tape squares of the original machine. Thus, the tape alphabet Z_i' of the new machine will consist of the ordered m-tuples of the original tape alphabet Z_i, that is,

$$[z_1, \ldots, z_m] \in Z_i' \quad \text{for all } z_1, \ldots, z_m \in Z_i$$

while the input alphabet is left unchanged. The internal states of the new machine will correspond to the internal states of the original machine but they will also represent the current positions of the read-write heads within the corresponding blocks. So they will have the form

$$[q, i_1, \ldots, i_k]$$

where q is the state of the original machine and i_j ($1 \leqslant i_j \leqslant m$) is the position of the j-th read-write head ($1 \leqslant j \leqslant k$) within the corresponding block. Each move of the original Turing machine can thus be simulated by an equivalent move of the new machine but the latter will use at most $S(n)/m$ tape squares from each of its work tapes. (The details of the construction are left to the reader.) Now, we can choose the value of m such that $1/m < c$ which gives the result. (We need at least one tape square even if $cS(n) < 1$.)

Corollary If L is in NSPACE($S(n)$), then for any $c > 0$, L is in NSPACE($cS(n)$). Similarly, for any $c > 0$, DSPACE($S(n)$) $\subseteq$ DSPACE($cS(n)$), because determinism will be preserved by the construction in Theorem 8.3.

Note that we generally assume that real valued complexity functions $f(n)$ represent, in fact, the values $\max(1, [f(n)])$ for all n. For $c > 1$ the above theorem and its corollary are trivial. In effect, the corollary can be formulated as

$$\text{NSPACE}(f(n)) = \text{NSPACE}(cf(n))$$

and

$$\text{DSPACE}(f(n)) = \text{DSPACE}(cf(n))$$

for any $c > 0$. These results are not surprising at all if we take into account the freedom of choosing arbitrarily large finite alphabets and arbitrarily large

---

* $[x]$ denotes here the integer part of x.

numbers of states and moves for the constructions of Turing machines. This freedom gives us the constant factor to the accuracy of the definition of complexity classes.

The problem becomes more delicate when we compare multitape and one-tape Turing machines, because the latter cannot have an extra input tape. But the construction used in Theorem 8.1 shows that the simulating one-tape machine would never use more than twice as many tape squares as the maximum number of those used by the original machine on any of its tapes. (The worst case arises when one read-write head is moving to the right while another is moving to the left all the time during the computation.) Hence, *for $f(n) > n$ the complexity classes NSPACE($f(n)$) and DSPACE($f(n)$) are independent of the number of tapes used in the definition of Turing machines.* (The factor 2 is irrelevant by Theorem 8.3.)

Corollary A language L is in the class NSPACE(cn) for some constant $c > 0$, iff L is context-sensitive. In other words, for any $c > 0$, NSPACE(cn) = $\mathcal{L}_1$.

Unfortunately, we do not know whether or not DSPACE(cn) forms a proper subclass of $\mathcal{L}_1$. The best result known so far is due to Savitch [1970], who has established a quadratic relationship between deterministic and nondeterministic space-complexity classes.

Definition 8.2 A function $S(n)$ is called space-constructible if there is a deterministic Turing machine which, for any given input of length n, would place a special marker symbol on the $S(n)$-th tape square of one of its tapes without using more than $S(n)$ squares on any tape.

Theorem 8.4 If L is in NSPACE($S(n)$) for some space-constructible function $S(n) \geqslant n$, then L is in DSPACE($S^2(n)$).

PROOF Let L be accepted by an $S(n)$ space-bounded nondeterministic Turing machine A, and let A have m states, k tapes, and at most z different symbols in each of its tape alphabets Z_i ($1 \leqslant i \leqslant k$). The number of possible configurations of A, when applied to any given input of length n, is bounded by

$$m(S(n))^k (z + 1)^{kS(n)}$$

where $(S(n))^k$ represents the number of possible positions of the k read-write heads and $(z + 1)^{S(n)}$ is the maximum number of possible contents of each tape. But, since $m \geqslant 1$ and $\log_2 S(n) < S(n)$, we have

$$m(S(n))^k (z + 1)^{kS(n)} < m^{S(n)} 2^{kS(n)} (z + 1)^{kS(n)} = a^{S(n)}$$

for $n \geqslant 1$ where a is independent of n. Now, if a word of length n is in L, then it is accepted by A with some sequence of at most $a^{S(n)}$ moves, for no configuration is repeated in the shortest sequence leading to an accepting configuration.

For a deterministic simulation of A we have to check whether the initial configuration C_0 can be reduced to a halting configuration C_h in at most $a^{S(n)}$ moves. Let $C_1 \overset{i}{\underset{A}{\Rightarrow}} C_2$ denote the fact that C_1 is reduced by A to C_2 in at most 2^i moves. Then $C_1 \overset{i}{\underset{A}{\Rightarrow}} C_2$ holds iff either $C_1 = C_2$ or $C_1 \underset{A}{\Rightarrow} C_2$ or else there is some configuration C_3 such that $C_1 \overset{i-1}{\underset{A}{\Rightarrow}} C_3$ and $C_3 \overset{i-1}{\underset{A}{\Rightarrow}} C_2$.

In view of Theorems 8.1 and 8.3 we can assume that A has only one tape. Then A can be simulated by a one-tape deterministic Turing machine A' as follows. The first part of the tape of A' will contain the input. The second part will be used to lay out $S(n)$ tape squares to measure the size of the so called frames. ($S(n)$ is space-constructible.) On the rest of the tape three frames will be laid out for every checking request of the form "is $C_1 \overset{i}{\underset{A}{\Rightarrow}} C_2$?", with C_1 and C_2 occupying two frames and i (which is at most $\log_2 a^{S(n)} = S(n)\log_2 a$) stored in the third frame.

First we place the initial configuration C_0 in the first frame. Then every possible configuration is generated separately in the second frame and checked whether it is a halting configuration. If so, then we have to check if $C_0 \overset{i}{\underset{A}{\Rightarrow}} C_h$ holds for $i = S(n)\log_2 a$. This we shall do recursively by checking if $C_0 \overset{i-1}{\underset{A}{\Rightarrow}} C'$ and $C' \overset{i-1}{\underset{A}{\Rightarrow}} C_h$ holds for some configuration C'. Each recursion step requires six new frames so we need at most $6i$ frames at any one time during the recursive evaluation of the relation $C_0 \overset{i}{\underset{A}{\Rightarrow}} C_h$. But each frame is of size $S(n)$, so we shall use at most $6(S(n))^2 \log_2 a$ tape squares where the constant factor is unimportant as we have seen before.

Corollary $\mathcal{L}_1 \subseteq \text{DSPACE}(n^2)$.

This corollary is just a special case of the theorem but it seems to be extremely difficult, if not impossible, to improve on this result. The time complexity of the simulation is already outrageous and it is questionable whether we can save much more space by further increasing the time consumption. The trade-off between the two basic resources, time and space, has also its theoretical limitations, but in general there is a great variety of compromising between the two extremes.

Example 8.1 Consider the language
$$L = \left\{ PcP^{-1} \mid P \in \{a, b\}^* \right\}$$

This context-free language can be recognized by a deterministic Turing machine with one work tape (in addition to the read only input tape) of space-complexity $\log_2 n$. This can be achieved by using the work tape as a binary counter. First we record the length of P on the work tape. (The details of designing a Turing machine to count in binary notation are left to the reader.) Then we start counting from $|P|$ down to 0 and for each value j of the counter we shall compare the j-th letter after the c with the j-th letter before the c. (This can be done by counting down from j to zero when going past the c and counting up again when coming back to c. The j-th letter before c can be found in a similar manner and the value of j can be restored by shifting back the input head to the center c.) Then the value of j will be decreased by one and a new comparison is performed unless the counter has been set to zero. This way we use only $\log_2 |P|$ squares from the work tape.

The same language, however, can be recognized much faster if we use $|P|$ squares from the work tape. In that case we can copy the first half of the input onto the work tape and then read the rest of the input while scanning the work tape backward to compare the corresponding letters. This second procedure requires only n moves, whereas the first one makes $4(1 + 2 + \cdots + (n-1)/2) = (n-1)(n+1)/2$ moves besides counting.

Let us now turn our attention to the time-complexity classes. First we prove the linear speed-up theorem which is the counterpart of Theorem 8.3.

Theorem 8.5 If L is accepted by a $T(n)$ time-bounded k-tape Turing machine with $k > 1$, then for any $c > 0$, L is accepted by a $cT(n)$ time-bounded k-tape Turing machine provided that $\inf_{n \to \infty} (T(n)/n) = \infty$.*

PROOF Each block of m adjacent tape squares of the $T(n)$ time-bounded Turing machine will be represented by one tape square in the simulating machine. The current positions of the read-write heads within these blocks can be recorded in the internal state of the simulating machine. Moreover, the internal state of the latter machine will store the contents of three adjacent blocks from each tape with the read-write head actually scanning the middle one of them. In order to obtain this information each head will be moved to the left once, to the right twice, and to the left once. Four steps of the simulating machine will thus be enough to perform internally every computation that does not leave these block-triples. The original

*For an infinite sequence of real numbers $\langle a_n \rangle$ the symbol $\inf_{n \to \infty} a_n$ denotes the greatest real number r such that for every $s < r$ there are at most finite many members in $\langle a_n \rangle$ which are smaller than s. If for any real number r there are only finite many members with $a_n < r$, then $\inf_{n \to \infty} a_n = \infty$.

machine has to make at least m moves before it can leave such a triple. The simulating machine will then again make four moves to change the contents of the triples and shift the read-write heads to the correct positions. (To the left once and to the right three times, or vice versa.) Altogether eight moves are enough to simulate at least m moves of the original machine.

At the beginning of the simulation the input word has to be read completely and encoded (m letters to one) on another tape. The read-write head should then be returned to the beginning of the new input word. Hence, the initialization would take $n + (n/m)$ moves. Thereafter, the actual simulation takes at most $(8/m)T(n)$ moves. Hence, the entire simulation takes at most

$$n + \frac{n}{m} + \frac{8}{m}T(n) = \frac{(m + 1)n + 8T(n)}{m}$$

moves. But, since $\inf_{n \to \infty} (T(n)/n) = \infty$, for every m there is some n_0 such that

$$(m + 1)n < T(n)$$

for all $n > n_0$. Now, we can choose m such that

$$\frac{9}{m} < c$$

and thus,

$$\frac{(m + 1)n + 8T(n)}{m} < \frac{9T(n)}{m} < cT(n)$$

for all $n > n_0$, which completes the proof.

Theorem 8.6 If L is accepted by a $T(n)$ time-bounded one-tape Turing machine, then for any $c > 0$, L is accepted by a $cT(n)$ time-bounded one-tape Turing machine, provided that $\inf_{n \to \infty} (T(n)/n^2) = \infty$.

PROOF The proof is similar to that of Theorem 8.5 only we need n^2 moves to encode the input word on the same tape.

To conclude this section we show that the one-tape simulation of a k-tape Turing machine will increase the number of moves at most to its square.

Theorem 8.7 If L is accepted by a $T(n)$ time-bounded k-tape Turing machine, then L is accepted by a $(T(n))^2$ time-bounded one-tape Turing machine, provided that $\inf_{n \to \infty} (T(n)/n^2) = \infty$.

PROOF This theorem follows directly from the construction used in Theorem 8.1, where each move of the k-tape machine has been simulated by a basic cycle shifting the head back and forth between the leftmost and the rightmost markers. These are at most $2T(n)$ squares apart from each other, hence, each move of the k-tape machine is simulated by at most $2T(n) + 2(k + 1)$ moves of the one-tape machine, since at most k back shifts are performed in the second part of the basic cycle. Hence, the entire simulation takes at most

$$T(n)(2T(n) + 2(k + 1)) = 2(T(n))^2 + 2(k + 1)T(n)$$

moves which is essentially the same as $(T(n))^2$. Now, by Theorem 8.6 we can speed up to

$$c2(T(n))^2 + c2(k + 1)T(n)$$

for any $c > 0$. But, for any c with $0 < c < \frac{1}{2}$ we have

$$\frac{2c(k + 1)}{1 - 2c} < T(n)$$

for sufficiently large n. Hence

$$c2(T(n))^2 + c2(k + 1)T(n) < (T(n))^2$$

which completes the proof.

The study of complexity classes is a very important and highly developed branch of theoretical computer science. It is interesting to note that the complexity measures based on Turing machines are not so far from practical applications as they appear to be. It can be shown that the so-called random access machine (which is a more realistic model of digital computers) is not drastically faster than a Turing machine. More precisely, the time needed by a Turing machine to simulate the operation of a random access machine can be bounded by a polynomial expression of the time used by the latter (see, e.g., Aho-Hopcroft-Ullman [1974]). This underscores the importance of the *polynomial time-complexity classes* which include, in fact, all practically recognizable languages as opposed to those whose recognition algorithms give rise to the so-called *exponential explosion*. If the time used by a random access machine is bounded by some polynomial of the input length then the same is true also for some Turing machine. (A polynomial of a polynomial is again a polynomial, and never an exponential.) The big question here is whether nondeterministic Turing machines operating in polynomial time can be simulated by deterministic ones also operating in polynomial time. This question has been open for many years now and seems to resist every effort made to its solution. More details on complexity classes can be found in many books, e.g., in Aho-Hopcroft-Ullman [1974] (in particular, Chapter 10) and in Hopcroft-Ullman [1979] (Chapters 12 and 13).

8.3 COMPLEXITY OF CONTEXT-FREE LANGUAGE RECOGNITION

By the corollary of Theorem 8.4 we already know that $\mathcal{L}_1 \subseteq \text{DSPACE}(n^2)$. It is easy to see that context-free languages can be recognized in linear space by deterministic Turing machines.

Theorem 8.8 $\mathcal{L}_2 \subseteq \text{DSPACE}(n)$.

PROOF Our proof is based on Theorem 6.5 where we have developed a nondeterministic pushdown automaton for every context-free language. This pushdown automaton can be simulated by a nondeterministic one-tape Turing machine of space-complexity n, because the length of the nonterminal string in the pushdown store, plus the length of the rest of the input string never exceeds the length of the whole input string. This is so because the machine represents the inverse of a grammar in Chomsky normal form. Now, in a Chomsky normal form grammar the number of steps in the derivation of a terminal word P is exactly $2|P| - 1$ (see Theorem 3.3). Hence, by Theorem 8.2 we can simulate this machine by a deterministic 3-tape Turing machine which uses at most $2|P| - 1$ squares on its second tape and at most $|P|$ squares on the other tapes which establishes the result. (By Theorems 8.1 and 8.3 this can be simulated also by a deterministic one-tape Turing machine of space-complexity n.)

Note that the time-complexity of this deterministic Turing machine has no trivial upper bounds except for some exponential function of $|P|$. In order to achieve better time-bounds we have to proceed differently.

Theorem 8.9 Every context-free language can be recognized by a deterministic 3-tape Turing machine of time-complexity n^3.

PROOF First we describe informally a deterministic algorithm for recognizing an arbitrary context-free language and then we will show its implementation via a 3-tape Turing machine.

For the informal description we shall use an example. Without loss of generality we may assume that the grammar generating the context-free language is given in Chomsky normal form. Consider the grammar

$$G = (\{S, A, B, C, D\}, \{a, b\}, S, F)$$

where the rules in F are

$$S \to AB \quad S \to CD \quad S \to CB \quad S \to SS$$
$$A \to BC \quad C \to DD \quad B \to SC \quad D \to BA$$
$$A \to a \quad B \to b \quad C \to b$$

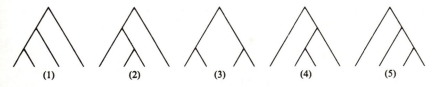

Figure 8.4 Derivation trees with four terminal nodes.

The deterministic algorithm will try to construct a derivation tree from the bottom up for every input word. Due to the Chomsky normal form, such a derivation tree—apart from the edges leading to the terminal nodes —must have the form of a binary tree. If, for instance, the input word has the length 4, then any of its derivation trees must have one of the five forms shown in Figure 8.4.

For a terminal word of length n the nodes of all possible derivation trees can be arranged in a triangular matrix whose last diagonal row has n cells and each other diagonal row has one less cell than does the row just below it. This matrix is called *recognition matrix* and its cells are subscripted as shown for $n = 5$ in Figure 8.5. It is easy to see that each derivation tree has precisely $2n - 1$ nodes out of the total of $(n/2)(n + 1)$ possible nodes.

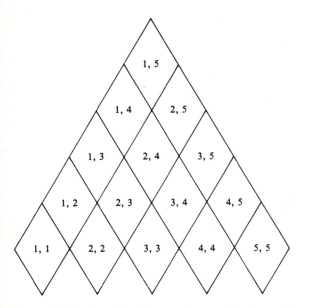

Figure 8.5 Subscripts of the elements of a recognition matrix with $n = 5$.

The aim of the recognizing algorithm is to enter the appropriate variables (nonterminal symbols) of the grammar in the cells of the matrix. The matrix will be filled up along the diagonals row by row, from the bottom up where each row is filled up from left to right. Given an input word $P = a_1 a_2 \cdots a_n$ with $a_i \in V_T$ then the bottom row will be computed in such a way that X is entered in the cell (i, i) whenever $X \to a_i$ is one of the rules of the grammar. (For the given example see Figure 8.6.) This means, of course, that each cell of the recognition matrix may contain zero or more variables. The next row will then be computed by entering X in the cell $(i, i + 1)$ whenever $X \to YZ$ is a rule of the grammar and Y occurs in the cell (i, i) while Z occurs in the cell $(i + 1, i + 1)$. In general, the cell (i, j) for $i < j$ will contain the variable X iff $X \to YZ$ is a rule of the grammar for some Y and Z such that Y occurs in the cell (i, k) while Z occurs in the cell $(k + 1, j)$ for some k with $i \leqslant k < j$ (see Figure 8.8).

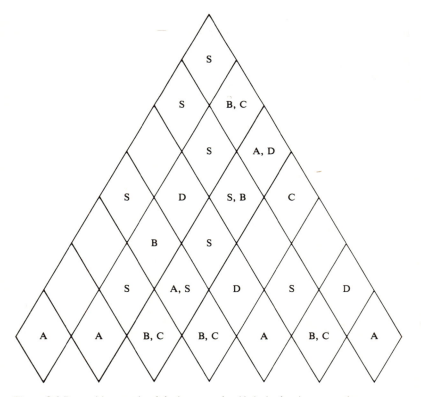

Figure 8.6 Recognition matrix of the input word *aabbaba* in the given example.

This computation makes certain that a variable X is entered in the cell (i, j) if and only if there exists a subtree representing the derivation of the portion of the input string below the cell (i, j), that is, iff

$$X \overset{*}{\underset{G}{\Rightarrow}} a_i \cdots a_j$$

Therefore, having finished the computation of the matrix, we have to check only whether S has been entered in the topmost cell indexed by $(1, n)$. If so, then the input word belongs to the language; otherwise it does not, since all possible ways of deriving the input word have been taken into account during the computation of the matrix.

The recognition matrix of the word *aabbaba* for the given grammar is shown in Figure 8.6 while the corresponding derivation tree is given in Figure 8.7.

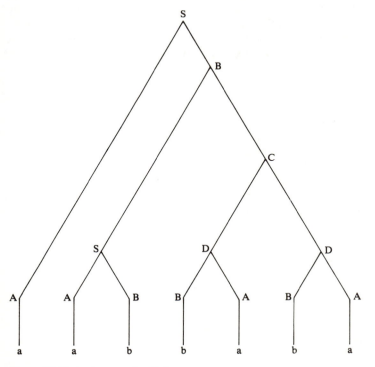

Figure 8.7 Derivation tree of *aabbaba*.

Let us consider now the realization of this recognizing algorithm in terms of a deterministic Turing machine. In order to minimize the running time of the algorithm we have to sacrifice extra space. Therefore, the entire recognition matrix will be stored twice so that each of the two worktapes will contain the whole matrix although in different layouts. The third tape will hold the input unchanged.

The basic cycle of the algorithm is the computation of the contents of the cell (i, j) using the pairs of cells (i, k) and $(k + 1, j)$ for $k = i, \ldots, j - 1$ as seen in Figure 8.8. Therefore, the processing speed can certainly be increased by storing the cells in the order of the orthogonal rows and columns, respectively (see Figure 8.9).

First the Turing machine would read the input word and initialize its work tapes by setting up as many frames for the cells of the recognition matrix as needed for the given input. (Each frame should be large enough to accommodate all possible symbols of V_N and we need exactly $n(n + 1)/2$ frames on each worktape.) Thus, the number of moves needed for the initialization is of the order of n^2.

Assume that at the beginning of the computation of the cell (i, j) the two read-write heads are scanning cells (i, i) and $(i + 1, j)$, respectively, on their tapes. Then in the basic cycle both heads will be shifted parallelly to the right until the first head reaches the cell $(i, j - 1)$. At this point the second head is scanning the cell (j, j). Then the first head moves one cell

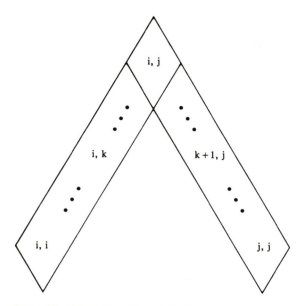

Figure 8.8 Computation of the cell (i, j).

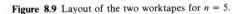

| 1, 1 | 1, 2 | 1, 3 | 1, 4 | 1, 5 | 2, 2 | 2, 3 | 2, 4 | 2, 5 | 3, 3 | 3, 4 | 3, 5 | 4, 4 | 4, 5 | 5, 5 |

Work tape 1

| 1, 1 | 1, 2 | 2, 2 | 1, 3 | 2, 3 | 3, 3 | 1, 4 | 2, 4 | 3, 4 | 4, 4 | 1, 5 | 2, 5 | 3, 5 | 4, 5 | 5, 5 |

Work tape 2

Figure 8.9 Layout of the two worktapes for $n = 5$.

to the right and stores the result in the cell (i, j). The second head has yet to be shifted to the left by at most n cells to reach the cell (i, j) on that tape. Altogether the computation of the cell (i, j) takes at most constant times $(n + n)$ moves. Now, the read-write heads should be replaced to the appropriate cells before the beginning of the next cycle. If $j < n$, then the next cell to be computed is $(i + 1, j + 1)$, therefore, both read-write heads should be shifted to the right by at most $n + 1$ cells to reach the cells $(i + 1, i + 1)$ and $(i + 2, j + 1)$, respectively. If $j = n$ but $i > 1$, then the next cell to be computed is $(1, n - i + 2)$; therefore, the first read-write head should be shifted back to the beginning of the tape while the second head should be shifted to the left as far as to reach the cell $(2, n - i + 2)$. In this case the number of moves taken to replace the heads is of the order $n(n + 1)/2$, but this happens only n times in the course of the entire computation.

Now, disregarding the constant factors, each cell is computed in $2n$ moves which gives $\frac{n}{2}(n + 1)2n = n^2(n + 1)$ moves. Replacing of the heads takes at most $\left(\frac{n}{2}(n + 1) - n\right)n + n\frac{n}{2}(n + 1) = n^3$ moves. The initialization took n^2 moves and thus the whole computation is of time-complexity n^3 which completes the proof.

The above algorithm, due independently to Cocke, Kazami, and Younger, is of space-complexity n^2 which is quite a reasonable trade-off in view of the fact that it is not known whether the time-bound n^3 could be significantly reduced for arbitrary context-free languages. By using the matrix multiplication algorithm of Strassen, the bound n^3 can be reduced asymptotically to $n^{2.81}$ (Valiant [1975]). Context-free languages occurring in practical applications usually belong to some subclass of $\mathcal{L}_2$, which can be recognized in linear time. Some of these subclasses will be studied later in Section 9.4.

8.4 THE HARDEST CONTEXT-FREE LANGUAGE

One way of solving a given problem is to reduce it to some other problem whose solution is already known. Such a reduction is best defined in terms of a mapping which converts every instance of the first problem into an instance of the other, and at the same time it allows for converting back each solution of the latter into a solution of the former. This method can also be used to find some upper bound to the complexity of a problem by adding together the complexity of the reduction itself and the complexity of the other problem to which the first one is reduced.

Assume that we are given two languages $L_1 \subseteq V_1^*$ and $L_2 \subseteq V_2^*$ and we have a mapping h from V_1^* into V_2^* such that *for every $P \in V_1^*$ its image $h(P)$ is in L_2 iff $P \in L_1$.* This means that $L_1 = h^{-1}(L_2)$ where $h^{-1}(L_2)$ denotes the image of L_2 under the inverse mapping h^{-1}. (In exact terms, $P \in h^{-1}(L_2)$ iff $h(P) \in L_2$.) Note that $h(h^{-1}(L_2)) \subseteq L_2$ where the inclusion may also be proper since not every word in L_2 must necessarily be in $h(V^*)$. Clearly, *the recognition of L_1 can be reduced to that of L_2, provided that the mapping h is computable*. Now, if h is a homomorphism, it can always be computed (letter by letter) in linear time and there is some constant c such that

$$|h(P)| \leqslant c|P|$$

for all $P \in V_1^*$. This computation can be implemented by a deterministic Turing machine with two tapes. Further, if L_2 can be recognized by a deterministic Turing machine of time-complexity $p(n)$ where $p(n)$ is some polynomial of n then L_1 can be recognized in $p(cn)$ time which is $p'(n)$ for some other polynomial with the same degree. This means that *the time-complexity of the recognition is preserved under inverse homomorphism for every polynomial time recognizable language*.

As we have seen before, context-free languages belong to the class $\text{DTIME}(n^3)$. Hence, the language $h^{-1}(L)$ is also in $\text{DTIME}(n^3)$ for every homomorphism h and context-free language L. Also, we know that for every context-free language L its homomorphic image $h(L)$ is in $\mathcal{L}_2$ for any homomorphism h (see Exercise 3.2). But, is it possible to find an appropriate homomorphism for each context-free language which would map it into a proper subset of $\mathcal{L}_2$? Interestingly enough, there exists a universal context-free language L_0 such that to every $L \in \mathcal{L}_2$ we can find a homomorphism h for which $L = h^{-1}(L_0)$. This language L_0 has been found by Greibach and is called the hardest context-free language because all other context-free languages have at most the same time-complexity as L_0 does (Greibach [1973]). Yet, the time-bound n^3 (or $n^{2.81}$ if you like) has not been improved for L_0. Nevertheless, it is worthwhile to know that any progress that can be made with respect to L_0 will be automatically extended to the whole class $\mathcal{L}_2$. The construction of the language L_0 will be described below.

First we define the so-called Dyck language which is a generalization of parenthesis systems.

Definition 8.3 The *Dyck language* over the $2n$-letter alphabet $V = \{e_1, \ldots, e_n, \bar{e}_1, \ldots, \bar{e}_n\}$ is the set of those words of V^* which can be reduced to the empty word λ by performing all possible cancellations of matching pairs in it, i.e., by substituting λ successively for every occurrence of the substrings $e_i \bar{e}_i$ $(i = 1, \ldots, n)$.

Note that the pairs $e_i \bar{e}_i$ are canceling each other only in that order. $(\bar{e}_i e_i \neq \lambda)$. It can be easily shown that for every word $P \in V^*$ these cancellations can be performed in any possible order without changing the result. Therefore, all words in V^* can be grouped into equivalence classes according to their remainders under cancellation. This means that P_1 and P_2 will be in the same class iff $\mu(P_1) = \mu(P_2)$ where $\mu(P)$ denotes the remainder of P under cancellation. Hence, the Dyck language is the equivalence class of words $P \in V^*$ with $\mu(P) = \lambda$.

The Dyck language is context-free as can be seen from the rules

$$S \to \lambda, S \to SS, \quad \text{and} \quad S \to e_i S \bar{e}_i \quad (i = 1, \ldots, n)$$

For the construction of L_0, we need only a 4-letter Dyck language that will be denoted by D. The letters of D will be used for encoding the rules of any context-free grammar. Three more letters are used as delimiters within the words of L_0.

Definition 8.4 Let D denote the Dyck language over the alphabet $\{e_1, e_2, \bar{e}_1, \bar{e}_2\}$ and let $V = \{e_1, e_2, \bar{e}_1, \bar{e}_2, b, c, d\}$ be a finite alphabet. Then the language $L_0 \subset V^*$ is defined as

$$L_0 = \{\lambda\} \cup \{X_1 c Y_1 c Z_1 d \cdots X_n c Y_n c Z_n d \mid n \geqslant 1, Y_1 \ldots Y_n \in \{b\} D,$$
$$X_1, \ldots, X_n, Z_1, \ldots, Z_n \in \{\lambda\} \cup (V - \{d\})^* \}$$

To show that L_0 is context-free we observe that the kernel of each word of L_0 is a word of D whose parts are randomly separated by the other components. We claim that L_0 can be generated by the following rules:

$S \to XbYcZd$	$Y \to cZdXY$	$Y \to YcZdX$
$Y \to \lambda$	$X \to Zc$	$Z \to e_1 Z$
$Y \to YY$	$Z \to \lambda$	$Z \to e_2 Z$
$Y \to e_1 Y \bar{e}_1$	$Z \to bZ$	$Z \to \bar{e}_1 Z$
$Y \to e_2 Y \bar{e}_2$	$Z \to cZ$	$Z \to \bar{e}_2 Z$

Theorem 8.10 To every context-free language L we can find a homomorphism h such that $L = h^{-1}(L_0)$.

PROOF Let G be a grammar in Greibach normal form generating L, if L is λ-free, or $L\text{-}\{\lambda\}$, if it is not. Let the variables of G be denoted by $A_1, \ldots, A_n$ where we can assume that the initial symbol A_1 does not occur on the right-hand sides of the rules. The rules of G will be encoded using the alphabet $\{e_1, e_2, \bar{e}_1, \bar{e}_2\}$ as follows:

To every rule of the form

$$f: A_1 \to aA_{q_1} \cdots A_{q_r} \qquad (r \geqslant 0)$$

the value

$$\varphi(f) = be_1 e_2^{q_r} e_1 \cdots e_1 e_2^{q_1} e_1$$

will be assigned. Further, to every rule of form

$$f: A_p \to aA_{q_1} \cdots A_{q_r} \qquad (p > 1, r \geqslant 0)$$

the value

$$\varphi(f) = \bar{e}_1 \bar{e}_2^p \bar{e}_1 e_1 e_2^{q_r} e_1 \cdots e_1 e_2^{q_1} e_1$$

will be assigned. The encoding function φ does not take care of the terminal letter occurring in the rule. But each rule contains precisely one terminal letter. So let

$$\{f_{a,1}, \ldots, f_{a,k_a}\}$$

be the set of all rules containing the terminal letter a. Then the homomorphic image of a will be defined as

$$h(a) = c\varphi(f_{a,1})c \cdots c\varphi(f_{a,k_a})d$$

Defining $h(a)$ similarly for all $a \in V_T$ the homomorphism h is defined for all P in V_T^* including $P = \lambda$. ($h(\lambda) = \lambda$.)

Now, we have to show that $L = h^{-1}(L_0)$. For this purpose we prove a lemma first.

Lemma Let G be in a Greibach normal form as above and let $a_{i_1}, \ldots, a_{i_k}$ be arbitrary terminal letters of G and $A_{j_1}, \ldots, A_{j_m}$ be arbitrary variables of G. Then the relation

$$A_1 \overset{*}{\underset{G}{\Rightarrow}} a_{i_1} \cdots a_{i_k} A_{j_1} \cdots A_{j_m} \qquad (m \geqslant 0)$$

holds iff $h(a_{i_1} \cdots a_{i_k})$ is of the form

$$h(a_{i_1} \cdots a_{i_k}) = X_1 c Y_1 c Z_1 d \cdots X_k c Y_k c Z_k d$$

where

$$Y_1 \cdots Y_k \in \text{HEAD}(\{b\}D)$$

and

$$\mu(Y_1 \cdots Y_k) = be_1 e_2^{j_m} e_1 \cdots e_1 e_2^{j_1} e_1$$

The HEAD of a language has been defined in Section 2.1. The lemma will be proved by induction on k.

Basis: Let $k = 1$ and let the relation

$$A_1 \underset{G}{\overset{*}{\Rightarrow}} a_{i_1} A_{j_1} \cdots A_{j_m} \qquad (m \geqslant 0)$$

be valid. This means that

$$A_1 \rightarrow a_{i_1} A_{j_1} \cdots A_{j_m} \qquad (m \geqslant 0)$$

must be a rule of G and thus $h(a_{i_1})$ has form

$$h(a_{i_1}) = X_1 c b e_1 e_2^{j_m} e_1 \cdots e_1 e_2^{j_1} e_1 c Z_1 d$$

which was to be shown.

Conversely, let us assume that

$$h(a_{i_1}) = X_1 c Y_1 c Z_1 d$$

with $Y_1 \in \mathrm{HEAD}(\langle b \rangle D)$ and $\mu(Y_1) = b e_1 e_2^{j_m} e_1 \cdots e_1 e_2^{j_1} e_1$. Then, by the definition of h, there must be a rule

$$A_1 \rightarrow a_{i_1} A_{j_1} \cdots A_{j_m}$$

in G which completes the proof of the basis of the induction.

Induction Step: Assume that the lemma is true for some $k \geqslant 1$. Assume further that the relation

$$A_1 \underset{G}{\overset{*}{\Rightarrow}} a_{i_1} \cdots a_{i_{k+1}} A_{j_1} \cdots A_{j_m} \qquad (m \geqslant 0)$$

holds. This derivation can be made leftmost in the form

$$A_1 \underset{G}{\overset{*}{\Rightarrow}} a_{i_1} \cdots a_{i_k} A_p A_{j_{r+1}} \cdots A_{j_m} \underset{G}{\Rightarrow} a_{i_1} \cdots a_{i_k} a_{i_{k+1}} A_{j_1} \cdots A_{j_m}$$

for some $p > 1$ and $0 \leqslant r \leqslant m$. (Of course, if $r = m$, then $A_{j_{r+1}}$ does not occur.) This means that the rule applied in the last step of the derivation has the form

$$A_p \rightarrow a_{i_{k+1}} A_{j_1} \cdots A_{j_r} \qquad (p > 1, 0 \leqslant r \leqslant m)$$

Hence,

$$h(a_{i_{k+1}}) = X_{k+1} c \bar{e} \bar{e}_2^p \bar{e}_1 e_1 e_2^{j_r} e_1 \cdots e_1 e_2^{j_1} e_1 c Z_{k+1} d$$

while the induction hypothesis gives us

$$h(a_{i_1} \cdots a_{i_k}) = X_1 c Y_1 c Z_1 d \cdots X_k c Y_k c Z_k d$$

with

$$\mu(Y_1 \cdots Y_k) = b e_1 e_2^{j_m} e_1 \cdots e_1 e_2^{j_{r+1}} e_1 e_1 e_2^p e_1$$

from which the first part of the assertion follows immediately.

Conversely, assume that

$$h(a_{i_1} \cdots a_{i_{k+1}}) = X_1 c Y_1 c Z_1 d \cdots X_k c Y_k c Z_k d X_{k+1} c Y_{k+1} c Z_{k+1} d$$

with

$$\mu(Y_1 \cdots Y_k Y_{k+1}) = be_1 e_2^{j_m} e_1 \cdots e_1 e_2^{j_1} e_1$$

for some $j_1, \ldots, j_m$. By the construction of h, Y_{k+1} can have only the form

$$Y_{k+1} = \bar{e}_1 \bar{e}_2^p \bar{e}_1 e_1 e_2^{q_r} e_1 \cdots e_1 e_2^{q_1} e_1$$

for some $p, q_1, \ldots, q_r$. This implies that

$$\mu(Y_1 \cdots Y_k) = be_1 e_2^{j_m} e_1 \cdots e_1 e_2^{j_{r+1}} e_1 e_1 e_2^p e_1$$

and $j_1 = q_1, \ldots, j_r = q_r$. Therefore, G must have a rule of the form

$$A_p \rightarrow a_{i_{k+1}} A_{j_1} \cdots A_{j_r}$$

while the induction hypothesis yields

$$A_1 \underset{G}{\overset{*}{\Rightarrow}} a_{i_1} \cdots a_{i_k} A_p A_{j_{r+1}} \cdots A_{j_m}$$

which implies that

$$A_1 \underset{G}{\overset{*}{\Rightarrow}} a_{i_1} \cdots a_{i_k} a_{i_{k+1}} A_{j_1} \cdots A_{j_m}$$

holds and this completes the proof of the lemma.

The proof of the theorem will now be simple. Let $P = a_{i_1} \cdots a_{i_k} \in V_T^*$. If $P \in L$ then

$$A_1 \underset{G}{\overset{*}{\Rightarrow}} a_{i_1} \cdots a_{i_k}$$

and thus, by applying the lemma with $m = 0$ we get

$$h(P) = X_1 c Y_1 c Z_1 d \cdots X_k c Y_k c Z_k d$$

with $Y_1 \cdots Y_k \in \text{HEAD}(\langle b \rangle D)$ and $\mu(Y_1 \cdots Y_k) = b$ which means that $h(P) \in L_0$. Conversely, if $h(P) \in L_0$ then there exists a derivation $A_1 \underset{G}{\overset{*}{\Rightarrow}} P$ hence, $P \in L$ which completes the proof.

CHAPTER NINE

SYNTAX ANALYSIS

9.1 THE CONNECTION BETWEEN SYNTAX AND SEMANTICS

So far we have studied the recognition (membership) problem of formal languages as a "yes-no" problem only. In practical applications, however, it is usually not enough to decide whether a given word belongs to some language but, if so, then it is also necessary to know how to derive that particular word in the given grammar. *Syntax analysis* or *parsing* is, therefore, concerned with the syntactical structure of the words belonging to the language. (Remember that a word in a formal language is more like a sentence in a natural language.) It should be emphasized that by definition the syntactical structure depends on the given grammar generating the language. Hence we are not totally free to convert the grammar into a weakly equivalent one being in some normal form (see Theorem 8.9) or otherwise more suitable for the parsing algorithm. Such changes could seriously affect the result of the ensuing syntax analysis of words and thereby their semantical interpretation.

For context-free grammars the syntactical structure is usually represented in the form of a derivation tree as described in Section 3.2. Formal language theory traditionally does not go beyond syntax, and thus semantical problems are not regarded as part of this theory. It should be clear, however, that the meaning of language constructs must be connected somehow to the underlying syntax. Therefore, any attempt at developing a reasonable theory of semantics should be based upon the corresponding theory of syntax. To illustrate this idea consider the following grammar defining the syntax of simple arithmetic expressions. As we have said before, the syntax of a language is usually identified with a grammar generating it.

Example 9.1 Consider the language of simple arithmetic expressions generated by the grammar

$$G = (\{S, A, B, C\}, \{+, *, (,), a, b\}, S, F)$$

where for the sake of simplicity a denotes any small letter from the English alphabet, b denotes any decimal digit, and the rules in F are the following:

$S \rightarrow S + A$	$A \rightarrow B$	$B \rightarrow Bb$
$S \rightarrow A$	$A \rightarrow C$	$B \rightarrow a$
$A \rightarrow A * B$	$A \rightarrow (S)$	$C \rightarrow Cb$
$A \rightarrow A * C$	$B \rightarrow Ba$	$C \rightarrow b$

Clearly, the strings derivable from B in this grammar will start with a letter and consist of letters and/or digits while the strings derivable from C consist of digits only. The semantics of arithmetic expressions are concerned with the meaning of such expressions.

A bottom-up approach would assign some meanings to the basic components first and then try to synthesize the meanings of the more complex structures in terms of the meanings of their constituent parts. This means that we have to define first the meanings of the symbols $+$, $*$, a, and b. But the specification of the operands related to the operations represented by $+$ and $*$ would depend already on the syntax (infix or prefix notation, precedence of operations, etc.) The operands represented by a string of digits will also have their meanings dependent on their structure, that is, on the order in which their digits appear when derived by the above grammar. The order of the execution of the operations appearing in some expression is again best defined with reference to the order of the application of the production rules in the course of the derivation of the expression. For instance, the expression

$$2 + 3 * 4$$

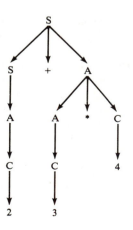

has a derivation tree which shows that $3 * 4$ forms a subexpression of $2 + 3 * 4$ so the multiplication is to be performed before the addition.

Another approach is to develop the meaning top-down by using the meanings (i.e., some kinds of general attributes) assigned to the grammatical categories represented by nonterminal symbols to derive the meanings of their components. This way we get the so called *inherited* attributes as opposed to the bottom-up development of the *synthesized* attributes. A combination of the two would be needed in most cases.

Similar ideas of syntax-directed translations are widely used in computer science for the construction of compilers (i.e., translators) for programming languages. It is, however, much harder to specify the semantics of a programming language than it is its syntax. For the time being, in contrast to syntax, no completely satisfactory theory of semantics is available that could be used for a sufficiently large variety of applications. Nevertheless, there are many significant results in this area and a large amount of research effort is spent incessantly on this subject. A systematic study of formal semantics is beyond the scope of this book but the interested reader can find some references in the bibliographic notes.

The point here is only to underscore the importance of the syntax analysis as the basis for the semantical evaluation of the words in the language. Also, it should be borne in mind that transformations of the grammars into weakly equivalent ones cannot be freely performed if semantic rules associated with the syntax are to be considered. Depending on the given rules one may also develop semantically equivalent transformations, but they will be certainly more delicate than weakly equivalent ones. It is, therefore, very important to have a general parsing algorithm that is efficient enough and works for an arbitrary context-free language without changing its syntax. Such an algorithm has been found by Earley [1970] and will be described in Section 9.3. But before presenting any parsing techniques, we would like to discuss the question of structural (that is, syntactical) ambiguity.

9.2 AMBIGUITY

A word $P \in L(G)$ may have more than one derivation in the grammar G. Two derivations which differ only in the order of the application of the rules (but not in the rules that are applied at the given places) are considered to be essentially the same. For context-free grammars this means that essentially different derivations are represented by different derivation trees as can be seen in this example:

$$G = (\{S\}, \{a, +, *\}, S, \{S \rightarrow S + S, S \rightarrow S * S, S \rightarrow a\})$$

In this grammar the word $a * a + a$ has two different derivation trees, as shown in Figure 9.1. Naturally, the leftmost derivations corresponding to these

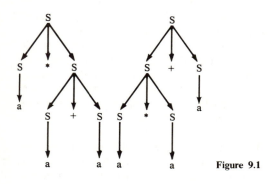

Figure 9.1

trees are also different:

$$S \Rightarrow S * S \Rightarrow a * S \Rightarrow a * S + S \Rightarrow a * a + S \Rightarrow a * a + a$$

and

$$S \Rightarrow S + S \Rightarrow S * S + S \Rightarrow a * S + S \Rightarrow a * a + S \Rightarrow a * a + a$$

These two derivations apply two different rules at the same (only) occurrence of S in the first step which means that they are essentially different. *For a context-free grammar G, a word P in $L(G)$ is said to be ambiguous iff P has two or more derivation trees (leftmost derivations) in G.*

A context-free grammar G is said to be ambiguous iff there exists a word in $L(G)$ which is ambiguous.

A context-free language L is said to be inherently ambiguous iff every context-free grammar generating L is ambiguous.

For every semantical evaluation, and also for every parsing algorithm, it would be desirable to know whether or not a given grammar is ambiguous. Unfortunately, the question of ambiguity for arbitrary context-free grammars is undecidable. We do not prove this fact here, for it is rather difficult. Also, we mention without proof that inherently ambiguous languages exist. Consider, for example, the context-free languages

$$L_1 = \{a^m b^m c^n \mid m, n \geqslant 1\}$$

and

$$L_2 = \{a^m b^n c^n \mid m, n \geqslant 1\}$$

Their union $L_1 \cup L_2$ is an inherently ambiguous context-free language because every context-free grammar which can be found for it would produce two different leftmost derivations for the words of form

$$a^n b^n c^n$$

belonging to both L_1 and L_2.

If we wish to have a decision procedure for testing for unambiguity, then we must restrict ourselves to appropriate subclasses of context-free grammars. In Section 9.4 we shall study some of those subclasses.

Regular languages can always be generated by unambiguous grammars, since each of them is recognized by a deterministic finite automaton which accepts each word of the language by a unique sequence of moves corresponding to a unique derivation in the related type 3 grammar. Consequently, no type 3 language is inherently ambiguous.

Now, we describe an interesting mathematical method for studying the ambiguity of context-free grammars. Let $G = (V_N, V_T, S, F)$ be a λ-free context-free grammar whose variables are denoted by $X_1, \ldots, X_n$ where $X_1 = S$. Assume further that G has no $X_i \rightarrow X_j$ type rules. For all $P \in V_T^*$ we define the integer number $d(P, G) \geqslant 0$ to be the *degree of ambiguity of P with respect to G*, which is the number of the different derivation trees of P. If $P \notin L(G)$ then obviously $d(G, P) = 0$. Consider now the *formal power series*

$$f(G) = \sum_{P \in V_T^*} d(P, G)P$$

where the summation is extended over the entire V_T^*. The proper meaning of formal addition and formal multiplication is immaterial for the formal treatment of these operations as long as they have the required properties. Addition is supposed to be a commutative and associative operation while multiplication is associative but not necessarily commutative. These operations are assumed to obey the distributive law and the constant factors are also treated in the usual way. With this much taken for granted, we can define the sum (term by term addition) and the product (Cauchy-product) of any two formal power series. Namely, let

$$h_1 = \sum_{i=1}^{\infty} a_i P_i$$

and

$$h_2 = \sum_{i=1}^{\infty} b_i P_i$$

be two arbitrary formal power series where P_i is in V_T^* whereas a_i and b_i are numeric factors for which addition and multiplication have their usual meaning. Then we adopt the definitions

$$h_1 + h_2 = \sum_{i=1}^{\infty} (a_i + b_i) P_i$$

and

$$h_1 h_2 = \sum_{i=1}^{\infty} \left(\sum_{P_j P_k = P_i} a_j b_k \right) P_i$$

The product $P_j P_k$ will now be defined as the catenation of P_j and P_k which makes good sense as we shall see next.

Our aim is to compute the formal power series $f(G)$ from the grammar G. To this end we compute a sequence of formal polynomials (finite power series) approximating $f(G)$. Let

$$P_{i,1}, \ldots, P_{i,m_i} \qquad (i = 1, \ldots, n)$$

denote the right-hand sides of those rules whose left-hand side is X_i. The formal polynomials f_i^k for $i = 1, \ldots, n$ and $k \geqslant 0$ will be defined recursively as follows:

$$f_i^0 = 0 \qquad (i = 1, \ldots, n)$$

and for $k > 0$

$$f_i^k = \sum_{j=1}^{m_i} F^{k-1}(P_{i,j}) \qquad (i = 1, \ldots, n)$$

where $F^{k-1}(P)$ denotes the polynomial obtained from the word P by substituting f_i^{k-1} for every occurrence of X_i $(i = 1, \ldots, n)$ in P. The resulting expression can be rearranged according to the properties of the formal operations.

It is easy to see that f_i^k will represent the sum of all terminal words which can be derived in G from X_i in at most k steps. Namely, for $k = 1$ we have

$$f_i^1 = F^0(P_{i,1}) + \cdots + F^0(P_{i,m_i}) \qquad (i = 1, \ldots, n)$$

and substituting zero for every nonterminal in $P_{i,j}$ will result in zero unless $P_{i,j}$ contains only terminal letters. By continuing this reasoning we get the result which also implies that

$$f(G) = \lim_{k \to \infty} f_1^k$$

Actually, in the approximating polynomial f_1^{2k-1} the coefficient of any word P with $|P| \leqslant k$ is already equal to $d(P, G)$. This follows from the observation that the derivation of P has at most $2k - 1$ steps if $|P| \leqslant k$. (Having neither $X_i \to \lambda$ nor $X_i \to X_j$ rules, every rule with nonterminal(s) on its right-hand side must be length increasing, and thus we can use such rules in at most $k - 1$ steps. On the other hand, we can use those rules which have terminal(s) on their right-hand sides k-times at most.) So we have a procedure to compute $d(P, G)$ for arbitrary P which can be used to decide whether or not P is ambiguous in G. It is, of course, much harder to decide whether the grammar G is ambiguous. For an unambiguous grammar G we must have $d(P, G) \leqslant 1$ for all P in V_T^*.

The connection with ordinary power series can be made closer by substituting z for all terminal letters in $f(G)$ which gives us z^k for every $P \in V_T^*$ with $|P| = k$. Thus, we get an ordinary power series of the form

$$f(z) = \sum_{k=0}^{\infty} D(k, G) z^k$$

where
$$D(k, G) = \sum_{|P| = k} d(P, G)$$

The values of $D(k, G)$ may also be computed from the equations obtained from the rules of G by substituting z for each terminal symbol; that is, from the equations

$$X_i = \sum_{j=1}^{m_i} P_{i,j}(z) \qquad (i = 1, \ldots, n)$$

Using these equations we can compute successively the coefficients of $f(z)$ as we did it with $f(G)$.

In some cases we can do even better by solving the above system of simultaneous equations for X_i, which yields an algebraic expression of z for each X_i. The algebraic expression obtained in this fashion for X_1 must be equal to the power series $f(z)$. Such a computation is given in the following example.

Example 9.2 Consider the grammar with the rules

$$X_1 \to aX_3 \quad X_1 \to bX_2 \qquad X_1 \to aX_3X_1 \quad X_1 \to bX_2X_1$$

$$X_2 \to a \quad X_2 \to bX_2X_2$$

$$X_3 \to b \quad X_3 \to aX_3X_3$$

It is easy to see that this grammar is weakly equivalent to the one given in Example 1.1. The approximating polynomials of $f(G)$ will be:

$$f_1^0 = f_2^0 = f_3^0 = 0$$

$$f_1^1 = 0$$

$$f_2^1 = a$$

$$f_3^1 = b$$

$$f_1^2 = af_3^1 + bf_2^1 + af_3^1f_1^1 + bf_2^1f_1^1 = ab + ba$$

$$f_2^2 = a + bf_2^1f_2^1 = a + baa$$

$$f_3^2 = b + af_3^1f_3^1 = b + abb$$

$$f_1^3 = af_3^2 + bf_2^2 + af_3^2f_1^2 + bf_2^2f_1^2 = a(b + abb) + b(a + baa)$$
$$+ a(b + abb)(ab + ba) + b(a + baa)(ab + ba)$$

$$\vdots$$

Next we show with the aid of $f(z)$ that this grammar is unambiguous. Consider the algebraic equations obtained from the rules:

$$X_1 = zX_3 + zX_2 + zX_3X_1 + zX_2X_1$$
$$X_2 = z + zX_2X_2$$
$$X_3 = z + zX_3X_3$$

The algebraic solution of this system of equations is

$$X_2 = \frac{1 - \sqrt{1 - 4z^2}}{2z} \qquad X_3 = \frac{1 - \sqrt{1 - 4z^2}}{2z}$$

$$X_1 = \frac{z(X_2 + X_3)}{1 - z(X_2 + X_3)} = (1 - 4z^2)^{-1/2} - 1$$

The last expression can be expanded into a power series using the formula of the binomial series

$$(1 - u)^{-1/2} = \sum_{k=0}^{\infty} (-1)^k \frac{\left(-\frac{1}{2}\right)\left(-\frac{3}{2}\right) \cdots \left(-\frac{2k-1}{2}\right)}{k!} u^k$$

$$= 1 + \sum_{k=1}^{\infty} \frac{1 \cdot 3 \cdots (2k-1)}{2^k k!} u^k = 1 + \sum_{k=1}^{\infty} \frac{(2k)!}{2^k k! 2^k k!} u^k$$

Hence we get

$$(1 - 4z^2)^{-1/2} - 1 = \sum_{k=1}^{\infty} \frac{(2k)!}{(k!)^2} z^{2k}$$

which means that

$$D(2k, G) = \frac{(2k)!}{(k!)^2}$$

But a simple combinatorial argument shows that the number of words in $L(G)$ containing precisely k occurrences of both a's and b's is equal to

$$\frac{(2k)!}{(k!)^2}$$

Hence, each of these words can have only one leftmost derivation in G because the equation

$$\sum_{|P|=2k} d(P, G) = \frac{(2k)!}{(k!)^2}$$

then implies $d(P, G) = 1$ for all $P \in L(G)$. Therefore, G is unambiguous, which was to be shown.

The difference between $D(k, G)$ and the number of words of length k belonging to $L(G)$ can be used in general to measure the degree of ambiguity of G.

Formal power series are useful tools in the theory of formal languages but one must be careful with the algebraic manipulations since multiplication is not commutative, that is, $X_i X_j \neq X_j X_i$ for $i \neq j$. Anyway, the algebraic solution of the system of equations related to the grammar is often prohibitive. For further information on formal power series see the bibliographic notes.

9.3 EARLEY'S ALGORITHM

In this section we present an efficient parsing algorithm for general context-free grammars. This algorithm due to Earley [1970] will also compute a recognition matrix for each input word, but unlike the recognition matrix of Theorem 8.9, the cells of the Earley matrix will contain production rules rather than nonterminals. The cells of this triangular matrix are supplied with indices as shown in Figure 9.2 for an input word of length n. Each cell will contain a set of *items* of the form

$$A \rightarrow X.Y$$

where $A \in V_N$, $X, Y \in (V_N \cup V_T)^*$, and $A \rightarrow XY$ is a production rule of the given grammar.

The position of the dot represents the progress made by the parsing process from left to right as visualized in Figure 9.3. This figure shows the situation after the first j letters of the input word have been processed. The

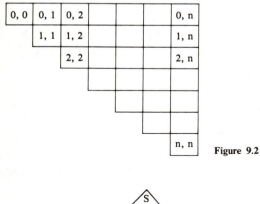

Figure 9.2

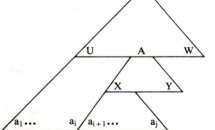

Figure 9.3

corresponding portions of the derivation tree are represented by triangles and we assume that we have found some words $U, W, X, Y \in (V_N \cup V_T)^*$ such that

$$S \overset{*}{\Rightarrow} UAW, \qquad A \rightarrow XY \in F$$

and $$U \overset{*}{\Rightarrow} a_1 \cdots a_i, \qquad X \overset{*}{\Rightarrow} a_{i+1} \cdots a_j$$

This fact will be recorded in the recognition matrix E by entering the item $A \rightarrow X.Y$ in the cell $E_{i,j}$. We do not know, however, whether the rest of the input word is derivable from YW. Therefore, we also include in $E_{i,j}$ all other items which may possibly be completed to get a derivation for the whole input word.

Given a context-free grammar $G = (V_N, V_T, S, F)$ and some word $P = a_1 \cdots a_n \in V_T^+$, the matrix E will be computed column by column from left to right with each column being computed bottom up except for the cells $E_{i,i}$ in the main diagonal which are to be computed at last in each column. (Note that the letters of P are indexed by their relative positions in P so a_i and a_j need not be different for $i \neq j$.) Initially the cells of E are supposed to be empty and they will be filled up by the following algorithm:

Step 1) Let $j = 0$. Include $S \rightarrow Z.U$ in $E_{0,0}$ for every $S \rightarrow ZU \in F$ with $Z \overset{*}{\Rightarrow} \lambda$.

Step 2) If $A \rightarrow X.ZBY \in E_{i,j}$ for some $i \leqslant j$ with $A, B \in V_N$ and $X, Z, Y \in (V_N \cup V_T)^*$ such that $Z \overset{*}{\Rightarrow} \lambda$ and $B \rightarrow VU \in F$ for some V and U with $V \overset{*}{\Rightarrow} \lambda$, then add $B \rightarrow V.U$ to $E_{j,j}$.

Step 3) If $j = n$, then the computation is finished; if $j < n$, then increase the value of j by 1 and let $i = j - 1$.

Step 4) If $A \rightarrow X.Za_jY \in E_{i,j-1}$ and $Z \overset{*}{\Rightarrow} \lambda$, then add $A \rightarrow XZa_j.Y$ to $E_{i,j}$.

Step 5) If $A \rightarrow X.ZBY \in E_{i,k}$ and $B \rightarrow U. \in E_{k,j}$ for some k with $i \leqslant k < j$ and $Z \overset{*}{\Rightarrow} \lambda$, then add $A \rightarrow XZB.Y$ to $E_{i,j}$.

Step 6) If $i \geqslant 1$, then decrease the value of i by 1 and return to step 4; else return to step 2.

Note that all Z here should be empty if the grammar is λ-free. Otherwise each symbol occurring in $Z \in V_N^*$ must belong to the set U as constructed in Theorem 3.1. (Then this construction forms a part of the algorithm but the set U should not be confused with a string denoted by U in this algorithm.)

In order to show that this algorithm works correctly for any context-free grammar we prove the following theorem.

Theorem 9.1 The recognition matrix E computed by Earley's algorithm for $P = a_1 \cdots a_n$ will contain the item $A \rightarrow X.Y$ in $E_{i,j}$ if and only if there exists some $W \in (V_N \cup V_T)^*$ such that

$$S \overset{*}{\Rightarrow} a_1 \cdots a_i A W$$

$$A \rightarrow XY \in F \quad \text{and} \quad X \overset{*}{\Rightarrow} a_{i+1} \cdots a_j$$

REMARK If $i = 0$, then the string $a_1 \cdots a_i$ should be interpreted as the empty string. Similarly $a_{i+1} \cdots a_j = \lambda$ for $i = j$.

PROOF First we show the *only if* part of the theorem. We use induction on the order of the computation of the items. When we compute the items for $E_{i,j}$ the dark cells shown in Figure 9.4 are already filled up. The order of computing the items within each cell will also be taken into account.

The items in $E_{0,0}$ are computed in step 1 and step 2. If $A \rightarrow X.Y \in E_{0,0}$ with $X \overset{*}{\Rightarrow} \lambda$ then either $A = S$ and $S \rightarrow XY \in F$ or $A \rightarrow XY \in F$ and $S \rightarrow .ZAW$ is already in $E_{0,0}$ for some $W \in (V_N \cup V_T)^*$ and $Z \overset{*}{\Rightarrow} \lambda$. The assertion is trivial in both cases.

Let $i < j$ and assume that the assertion is true for all $E_{r,s}$ with $s < j$ or $s = j$ and $i < r < j$. The items of $E_{i,j}$ are computed in steps 4 and 5. In step 4 we have $X = X'Za_j$ and $A \rightarrow X'.Za_jY \in E_{i,j-1}$ for some $X' \in$

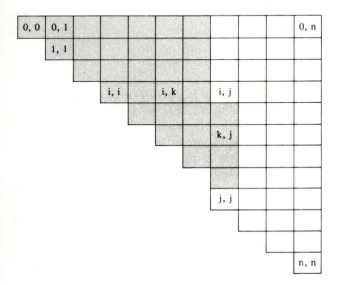

Figure 9.4

$(V_N \cup V_T)^*$ and $Z \overset{*}{\Rightarrow} \lambda$, so by the induction hypothesis

$$S \overset{*}{\Rightarrow} a_1 \cdots a_i A W \quad \text{and} \quad X' \overset{*}{\Rightarrow} a_{i+1} \cdots a_{j-1}$$

But then the assertion is clearly true for the corresponding $A \to X.Y \in E_{i,j}$. On the other hand, if this item is added to $E_{i,j}$ in step 5, then we have some k with $i \leqslant k < j$ such that $A \to X'.ZBY \in E_{i,k}$ and $B \to U. \in E_{k,j}$ with $X = X'ZB$ and $Z \overset{*}{\Rightarrow} \lambda$. Hence, by the induction hypothesis

$$S \overset{*}{\Rightarrow} a_1 \cdots a_i A W_1, \qquad X' \overset{*}{\Rightarrow} a_{i+1} \cdots a_k$$

and
$$S \overset{*}{\Rightarrow} a_1 \cdots a_k B W_2, \qquad U \overset{*}{\Rightarrow} a_{k+1} \cdots a_j$$

which imply that

$$X = X'ZB \overset{*}{\Rightarrow} a_{i+1} \cdots a_k a_{k+1} \cdots a_j$$

as required.

The items for $E_{j,j}$ are computed only in step 2 after having finished the computations for all $E_{i,j}$ with $i < j$. For each $B \to V.U$ added to $E_{j,j}$ we must have had some $i \leqslant j$ and $A \to X.ZBY$ in $E_{i,j}$ before. (At first $i < j$ must be the case.) Therefore, the induction hypothesis gives us

$$S \overset{*}{\Rightarrow} a_1 \cdots a_i A W \quad \text{and} \quad X \overset{*}{\Rightarrow} a_{i+1} \cdots a_j$$

But then $A \to XZBY \in F$ and $Z \overset{*}{\Rightarrow} \lambda$ imply

$$S \overset{*}{\Rightarrow} a_1 \cdots a_j B Y$$

which completes this part of the proof.

The *if* part of the theorem will be proved also by induction. Consider first $E_{0,0}$. Assume that

$$S \overset{*}{\Rightarrow} A W \quad \text{and} \quad A \to XY \in F$$

for some $A \in V_N$ and $X, Y, W \in (V_N \cup V_T)^*$. We use another induction on the number of steps in the derivation $S \overset{*}{\Rightarrow} AW$. For zero steps we have $A = S$ and $W = \lambda$, hence the item $A \to X.Y$ will certainly be included in $E_{0,0}$ by step 1. Assume that the assertion is true for such derivations with at most k steps and let $S \overset{*}{\Rightarrow} AW$ have $k + 1$ steps. Then clearly we have some $A' \in V_N$ and $W', W'' \in (V_N \cup V_T)^*$ such that the derivation

$$S \overset{*}{\Rightarrow} A'W'$$

has at most k steps and

$$A' \to ZAW'' \in F \quad \text{with } W = W''W' \quad \text{and} \quad Z \stackrel{*}{\Rightarrow} \lambda$$

By the induction hypothesis

$$A' \to .ZAW''$$

must be in $E_{0,0}$ and thus, $A \to X.Y$ will also be included in $E_{0,0}$ by step 2.

Assume now that the assertion is true for all $E_{r,s}$ with $s < j$, or $s = j$ and $i < r < j$. Assume further that

$$S \stackrel{*}{\Rightarrow} a_1 \cdots a_i A W$$

$$A \to XY \in F \quad \text{and} \quad X \stackrel{*}{\Rightarrow} a_{i+1} \cdots a_j$$

This implies that there is some $X' \in (V_N \cup V_T)^*$, and $Z \stackrel{*}{\Rightarrow} \lambda$ such that either

$$X = X'Za_j \quad \text{and} \quad X' \stackrel{*}{\Rightarrow} a_{i+1} \cdots a_{j-1}$$

or

$$X = X'ZB \quad \text{for some } B \in V_N$$

with

$$X' \stackrel{*}{\Rightarrow} a_{i+1} \cdots a_k, \quad \text{and} \quad B \stackrel{*}{\Rightarrow} a_{k+1} \cdots a_j$$

Then by the induction hypothesis either

$$A \to X'.Za_jY \in E_{i,\,j-1}$$

or else

$$A \to X'.ZBY \in E_{i,\,k}$$

with

$$S \stackrel{*}{\Rightarrow} a_1 \cdots a_k BY$$

In the first case

$$A \to X'Za_j.Y$$

will be included in $E_{i,\,j}$ by step 4 whereas in the second case there must be some rule $B \to U$ in F with $U \stackrel{*}{\Rightarrow} a_{k+1} \cdots a_j$. Hence, again by the induction hypothesis $B \to U. \in E_{k,\,j}$, and thus

$$A \to X'ZB.Y$$

will be included in $E_{i,\,j}$ by step 5.

Finally we have to consider the cell $E_{j,\,j}$ for $j > 0$. The steps of the derivation

$$S \stackrel{*}{\Rightarrow} a_1 \cdots a_j A W$$

can be arranged in such a way that its first part has the form

$$S \overset{*}{\Rightarrow} a_1 \cdots a_i A'W' \Rightarrow a_1 \cdots a_i X'AY'W'$$

with $\qquad X' \overset{*}{\Rightarrow} a_{i+1} \cdots a_j$ and $W = Y'W'$

If $i < j$, then by the induction hypothesis we have

$$A' \to X'.AY' \in E_{i,j}$$

and thus, by step 2 we get $A \to .XY \in E_{j,j}$. If $i = j$, then we repeat the reasoning with A' in place of A until we get $i < j$. Then we go back on this sequence using the induction hypothesis in each step, and this completes the proof.

As a corollary of Theorem 9.1 we get immediately the result that

$$P \in L(G) \quad \text{iff } S \to X.Y \in E_{0,n}$$

with $Y \overset{*}{\Rightarrow} \lambda$. The derivation tree or trees of P can be easily constructed from the matrix E if we store the references to the originating items along with each new item created by the process. (Three indices would do if items are numbered within each cell.) The space requirement of Earley's algorithm is obviously proportional to n^2. The time requirement is at most of the order n^3—when using a random access machine—but for many practical grammars the algorithm works much faster. Incidentally, λ-free grammars do not have great advantages with this algorithm. Also, ambiguous grammars can be treated by the algorithm without difficulty.

Example 9.3 Consider the context-free grammar

$$G = (\{S, A, B\}, \{a, +, *, (,)\}, S, F)$$

where the rules in F are as follows:

$S \to S + A$	$A \to A*B$	$B \to (S)$
$S \to A$	$A \to B$	$B \to a$

Let us compute the Earley matrix for the input word $a*a + a$. Two items

$$S \to .S + A$$

$$S \to .A$$

will be entered in $E_{0,0}$ by step 1. Then the items

$$A \to .A*B$$

$$A \to .B$$

$$B \to .(S)$$

$$B \to .a$$

will be added to $E_{0,0}$ by Step 2. After that $j = 1$ and $i = 0$ will be set by Step 3 and

$$B \rightarrow a.$$

will be entered in $E_{0,1}$ by step 4. In step 5 the items

$A \rightarrow B.$	(since $A \rightarrow .B \in E_{0,0}$ and $B \rightarrow a. \in E_{0,1}$)
$A \rightarrow A.*B$	(since $A \rightarrow .A*B \in E_{0,0}$ and $A \rightarrow B. \in E_{0,1}$)
$S \rightarrow A.$	(since $S \rightarrow .A \in E_{0,0}$ and $A \rightarrow B. \in E_{0,1}$)
$S \rightarrow S.+A$	(since $S \rightarrow .S + A \in E_{0,0}$ and $S \rightarrow A. \in E_{0,1}$)

will be added to $E_{0,1}$. (Note that step 5 and step 2 are closure operations that should be continued as long as they can produce new items.) Thereafter, we return to step 2 in order to compute $E_{1,1}$. But there is no item in $E_{0,1}$ to which an appropriate rule could be found in F, since no variable occurs in $E_{0,1}$ next to a dot. So the cell $E_{1,1}$ remains empty. Now step 4 is performed with $j = 2$ and $i = 1$, leaving $E_{1,2}$ empty (and it remains empty also in step 5). Then step 6 gives $i = 0$ and we return to step 4 which yields

$$A \rightarrow A*.B$$

for $E_{0,2}$. Step 5 adds nothing to $E_{0,2}$. Then step 2 enters

$$B \rightarrow .(S) \quad \text{and} \quad B \rightarrow .a$$

in $E_{2,2}$ due to $A \rightarrow A*.B \in E_{0,2}$.

S→.S+A S→.A A→.A*B A→.B B→.(S) B→.a	S→S.+A S→A. A→A.*B A→B. B→a.	A→A*.B	A→A*B. S→S.+B S→A. A→A.*B A→B.	S→S+.A	S→S+A.
	∅	∅	∅	∅	∅
		B→.(S) B→.a	B→a.	∅	∅
			∅	∅	∅
				A→.A*B A→.B B→.(S) B→.a	A→A.*B A→B. B→a.

Figure 9.5 The Earley matrix of $a*a + a$.

The rest of the computation will not be detailed here but the completed matrix is given in Figure 9.5 on page 149. As we have here

$$S \to S + A. \in E_{0,5}$$

it follows from Theorem 9.1 that

$$S \to S + A \in F \quad \text{and} \quad S + A \overset{*}{\Rightarrow} a^*a + a$$

so the input word belongs to the language $L(G)$. (The computation of $E_{5,5}$ is not necessary.) The items can be represented in a computer by pairs of integers, the first giving the number of the rule and the second giving the position of the dot.

9.4 $LL(K)$ AND $LR(K)$ GRAMMARS

In this section we discuss briefly some of the useful subclasses of context-free grammars which have fast parsing algorithms. The most common parsing algorithms fall into two categories: *top down* or *bottom up*, meaning the way of constructing the derivation tree for a given input word. The input word is always supposed to be read from left to right, therefore, the step-by-step development of derivation trees represents *leftmost derivations* in case of top-down construction whereas it produces a *rightmost derivation in reverse sense* when bottom-up construction is used.

Earley's algorithm of the previous section is a top-down construction where we carry the lists of all items (representing, in fact, a developing forest) which are candidates at that point for being applied later when more of the input string is seen. The power of this algorithm is due precisely to this parallel handling of those alternatives which are still open at any given point. Otherwise we may use trial and error methods where backtracking is needed after each unsuccessful trial.

The optimal solution would be to find the correct production rule (leading to the derivation of the actual input string) in every step of the parsing process so that the construction of the parsing tree proceeds in a strictly deterministic and sequential manner with no backtracking at all. But how can we predict the correct production rule when more than one is applicable and we have not seen the rest of the input word yet? Such a prediction is impossible in general, but we can find important subclasses of grammars where a limited look ahead is enough to make correct decisions at any given point of the parsing process.

In top-down parsing we produce intermediate strings U (sometimes called sentential forms due to the analogy with those of a natural language) for which we know that $S \overset{*}{\Rightarrow} U$ holds but do not know yet whether $U \overset{*}{\Rightarrow} P$ where P is our input. Bottom-up parsers, on the other hand, work with intermediate strings U such that $U \overset{*}{\Rightarrow} P$, but it is not yet known whether $S \overset{*}{\Rightarrow} U$ is true.

Let us consider the top-down approach first. Suppose we have a context-free grammar $G = (V_N, V_T, S, F)$ and a word $P \in V_T^*$ for which we have already produced the initial part of a leftmost derivation

$$S \Rightarrow U_1 \Rightarrow \cdots \Rightarrow U_i$$

where $$U_i = XAW \quad \text{and} \quad P = XY$$

for some $X, Y \in V_T^*$, $A \in V_N$, and $W \in (V_N \cup V_T)^*$. The next step in the leftmost derivation must be the expansion of A using some rule

$$A \rightarrow Q \in F$$

If there are several productions in F with A on the left-hand side, then the choice should be based on the information recorded in W as a result of the process so far and on the information about Y, the rest of the input string. A context-free grammar is called $LL(k)$, *for left to right scan, producing a leftmost derivation, with k symbol lookahead*, if we can always make a correct decision by examining at most the first k symbols of Y. (If the choice can be made without consulting Y then the grammar is $LL(0)$ and generates at most one word.) $LL(1)$ grammars have been used successfully for constructing compilers for many programming languages.

An $LL(1)$ grammar can be easily converted into a deterministic pushdown automaton that accepts the corresponding language. For this purpose we use the arrangement shown in Figure 9.6. Initially, the symbol S alone is stored in the stack. Each time we find a nonterminal A on top of the stack we replace it by the right-hand side of the corresponding rule $A \rightarrow Q$ so we get QW in the stack. If the top of the stack is a terminal symbol then it will be compared with the next input symbol, i.e., with the first letter of Y and if they match then it will be popped from the stack while the input head is shifted to the right to read the next input symbol.

It is interesting to note that the Greibach normal form seems to lend itself for $LL(1)$ processing. Namely, the first letter of Q must be a terminal symbol and thus the $LL(1)$ condition is obviously satisfied if for each variable A the

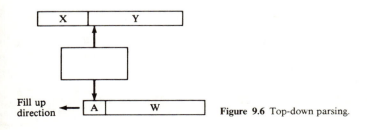

Fill up direction

Figure 9.6 Top-down parsing.

corresponding set of rules

$$A \to Q_1, \ldots, A \to Q_m$$

has the property that the first letter of Q_i is different from that of Q_j whenever $i \neq j$. If this condition fails, we may or may not be able to make a correct decision by using more information on W and Y. The precise definition of the $LL(k)$ condition will be given below. The formal definition makes use of the notation $P : k$ meaning the k letter long initial segment of P, that is,

$$P : k = \begin{cases} X, & \text{if } P = XY \text{ for some } X, Y \text{ with } |X| = k \\ P, & \text{if } |P| < k \end{cases}$$

Definition 9.1 A context-free grammar is $LL(k)$ if and only if for every pair of leftmost derivations of the form

$$S \overset{*}{\Rightarrow} XAW \Rightarrow XQW \overset{*}{\Rightarrow} XY \quad \text{and} \quad S \overset{*}{\Rightarrow} XAW \Rightarrow XRW \overset{*}{\Rightarrow} XZ$$

with $X, Y, Z \in V_T^*$ the equality $Y : k = Z : k$ implies $Q = R$.

A context-free language is called $LL(k)$ if it is generated by an $LL(k)$ grammar.

Theorem 9.2 Every $LL(k)$ grammar is nonambiguous.

PROOF The assertion follows immediately from Definition 9.1 since an ambiguous grammar has more than one leftmost derivation for some word(s). But then the $LL(k)$ condition is obviously violated for two derivations of the same word.

Example 9.4 Consider the grammar

$$G = (\{S, A, B\}, \{a, b\}, S, F)$$

with the rules in F

$$\begin{array}{lll} S \to aAB & A \to a & A \to bS \\ S \to bBA & B \to b & B \to aS \end{array}$$

It can be shown that this grammar is $LL(1)$. Indeed, let us be given two leftmost derivations of the form

$$S \overset{*}{\Rightarrow} XSW \Rightarrow XQW \overset{*}{\Rightarrow} XY \quad \text{and} \quad S \overset{*}{\Rightarrow} XSW \Rightarrow XRW \overset{*}{\Rightarrow} XZ$$

with $Y : 1 = Z : 1$. Then two cases arise, namely, either

$$Y : 1 = Z : 1 = a, \quad \text{which implies} \quad R = Q = aAB$$

or $\qquad Y : 1 = Z : 1 = b, \quad \text{which implies} \quad R = Q = bBA$

Similarly, for leftmost derivations of the form

$$S \overset{*}{\Rightarrow} XAW \Rightarrow XQW \overset{*}{\Rightarrow} XY \quad \text{and} \quad S \overset{*}{\Rightarrow} XAW \Rightarrow XRW \overset{*}{\Rightarrow} XZ$$

$Y:1 = Z:1 = a$ implies $Q = R = a$ and $Y:1 = Z:1 = b$ implies $Q = R = bS$. Finally, for

$$S \stackrel{*}{\Rightarrow} XBW \Rightarrow XQW \stackrel{*}{\Rightarrow} XY$$

$$S \stackrel{*}{\Rightarrow} XBW \Rightarrow XRW \stackrel{*}{\Rightarrow} XZ$$

$Y:1 = Z:1 = a$ implies $Q = R = aS$, while $Y:1 = Z:1 = b$ implies $Q = R = b$.

Example 9.5 The grammar $G = (\{S, A, B\}, \{a, b, c\}, S, F)$ with the rules

$S \to A$	$A \to aAb$	$B \to aBc$
$S \to B$	$A \to ab$	$B \to ac$

is not $LL(k)$ for any k. That is, for every k we can find leftmost derivations of form

$$S \stackrel{*}{\Rightarrow} S \Rightarrow A \stackrel{*}{\Rightarrow} a^k b^k \quad \text{and} \quad S \stackrel{*}{\Rightarrow} S \Rightarrow B \stackrel{*}{\Rightarrow} a^k c^k$$

where $a^k b^k : k = a^k c^k : k = a^k$ but $A \neq B$.

Example 9.6 Consider the grammar

$$G = (\{S\}, \{a, b\}, S, \{S \to aaSb, S \to ab, S \to bb\})$$

This grammar generates the language

$$L(G) = \{a^{2m+1} b^{m+1} | m \geqslant 0\} \cup \{a^{2m} b^{m+2} | m \geqslant 0\}$$

For a top-down parser it is easy to decide which rule is to be applied by looking ahead two symbols: if two a's follow then the first, if ab follows then the second, and if bb follows the third one. So this grammar is $LL(2)$ but obviously not $LL(1)$. However, the same language can be generated by an $LL(1)$ grammar with the rules

$S \to aA$	$A \to aSb$
$S \to bb$	$A \to b$

For further analysis of context-free grammars we define the FIRST sets for $W \in (V_N \cup V_T)^*$ as

$$\text{FIRST}_k^G(W) = \left\{ P : k \,|\, W \stackrel{*}{\underset{G}{\Rightarrow}} P \text{ and } P \in V_T^* \right\}$$

where G can be omitted if it is implied by the context. Similarly, we define the FOLLOW sets as

$$\text{FOLLOW}_k^G(W) = \left\{ \text{FIRST}_k^G(Y) | S \stackrel{+}{\Rightarrow} XWY \text{ for some } X, Y \in (V_N \cup V_T)^* \right\}$$

Theorem 9.3 A context-free grammar is $LL(k)$ if and only if the following condition holds:

If $A \to Q$ and $A \to R$ are two different rules with A on the left-hand side and there is a leftmost derivation $S \overset{*}{\Rightarrow} XAW$ for some $X \in V_T^*$ and $W \in (V_N \cup V_T)^*$, then

$$\text{FIRST}_k(QW) \cap \text{FIRST}_k(RW) = \varnothing$$

PROOF This theorem follows immediately from Definition 9.1; in fact, it is only a reformulation thereof.

REMARK A context-free grammar is called nonredundant if each of its variables occurs in the derivation of some terminal word. (See Definition 3.4.)

Thus, a nonredundant context-free grammar is $LL(1)$ iff for every variable A the set of the corresponding rules

$$\{A \to Q_1, \ldots, A \to Q_n\}$$

satisfies these two conditions:

1) $\text{FIRST}_1(Q_1) \cap \text{FIRST}_1(Q_j) = \varnothing$ for $i \neq j$
2) If $Q_i \overset{*}{\Rightarrow} \lambda$ then for all $j \neq i$ $(1 \leqslant j \leqslant n)$

$$\text{FIRST}_1(Q_j) \cap \text{FOLLOW}_1(A) = \varnothing$$

Thus we can use the FIRST and FOLLOW sets to check whether a grammar is $LL(k)$. On the other hand, it can be shown that *left recursion* is incompatible with the $LL(k)$ property for any k. By left recursion we mean here the existence of a derivation of the form

$$A \overset{+}{\Rightarrow} AP$$

for some $A \in V_N$ and $P \in (V_N \cup V_T)^*$. The relation $\overset{+}{\Rightarrow}$ denotes a derivation with at least one step. A variable for which such a derivation exists is called *left recursive*. Two variables A and B are called *mutually left recursive* if there are derivations of form

$$A \overset{+}{\Rightarrow} BP_1 \quad \text{and} \quad B \overset{+}{\Rightarrow} AP_2$$

for some $P_1, P_2 \in (V_N \cup V_T)^*$. The mutual left recursiveness is clearly a symmetrical, reflexive, and transitive relation. Thus the left recursive variables can be divided into equivalence classes which will be used in the next theorem.

Theorem 9.4 If a nonredundant context-free grammar has a left recursive variable then it is not $LL(k)$ for any k.

PROOF Assume that there is a left recursive variable in G and consider the class of mutually left recursive variables containing this variable. In this class there must be a variable A for which the right-hand side of at least one of the corresponding rules begins with some letter not belonging to this class. (Otherwise we could never get rid of the variables of this class in a derivation which makes use of one of them.) Therefore, we have at least two different rules for A, say,

$$A \rightarrow Q \quad \text{and} \quad A \rightarrow R$$

where the first is left recursive and the second is not. This means that $Q \overset{*}{\Rightarrow} AY$ for some $Y \in (V_N \cup V_T)^*$ but $R \overset{*}{\Rightarrow} AZ$ does not hold for any Z. Then for every leftmost derivation of the form

$$Q \overset{*}{\Rightarrow} AY \Rightarrow RY$$

two cases arise:

Case 1) $Y \overset{*}{\Rightarrow} \lambda$. Since the grammar is nonredundant, there must be a leftmost derivation of the form

$$S \overset{*}{\Rightarrow} XAW \Rightarrow XRW \overset{*}{\Rightarrow} P$$

where P is in V_T^*. But then $Q \overset{*}{\Rightarrow} R$ and thus, $QW \overset{*}{\Rightarrow} RW$ which implies

$$\varnothing \neq \text{FIRST}_k(RW) \subseteq \text{FIRST}_k(QW)$$

in contrast with the condition of Theorem 9.3.

Case 2) $Y \overset{*}{\not\Rightarrow} \lambda$. Now there is some $P \in V_T^+$ such that $Y \overset{*}{\Rightarrow} P$. Further, let $R \overset{*}{\Rightarrow} U$ for some $U \in V_T^*$ and choose n such that $|U| + n|P| > k$, that is, $|UP^n| > k$. Hence

$$UP^n : k \in \text{FIRST}_k(RP^n)$$

and from $QP^n \overset{*}{\Rightarrow} RP^{n+1}$ we get

$$UP^n : k = UP^{n+1} : k \in \text{FIRST}_k(QP^n)$$

Finally, since

$$S \overset{*}{\Rightarrow} XAW \overset{*}{\Rightarrow} XAP^nW$$

and

$$UP^n : k \in \text{FIRST}_k(RP^n) \cap \text{FIRST}_k(QP^n)$$

the condition of Theorem 9.3 cannot be satisfied, which completes the proof.

Left recursion can surely be eliminated by the Greibach normal form but it does not necessarily result in an $LL(k)$ grammar for some k. More improvement can be achieved by *left factoring* the rules of the form

$$A \to XP_1, \ldots, A \to XP_n$$

which means replacing them by

$$A \to XB$$

and
$$B \to P_1, \ldots, B \to P_n$$

This way we can eliminate the identical initial parts from the right-hand sides of related rules, but left recursion remains. (Left recursion may reappear even if we start with the Greibach normal form.) In short, there is no general method to transform every context-free grammar into an equivalent $LL(k)$ one. Moreover, for all $k \geqslant 0$ there are $LL(k + 1)$ grammars which are not equivalent to any $LL(k)$ grammar. For a proof of this we refer to the literature; in particular, see Kurki-Suonio [1969].

LF, for *left factored*, languages have been defined by Wood [1969] in the following way: For every rule $A \to Q$ we define its left terminal set, denoted by $[A, Q]$, as

$$[A, Q] = \left\{ Y : 1 \,|\, \text{there are words } X, Y \in V_T^* \text{ and } W \in (V_N \cup V_T)^* \text{ such that} \right.$$
$$\left. S \overset{*}{\Rightarrow} XAW \text{ and } AW \Rightarrow QW \overset{*}{\Rightarrow} Y \right\}$$

A variable A is called left factored if for each pair of rules with A on the left-hand side

$$A \to Q_1 \quad \text{and} \quad A \to Q_2$$

with $Q_1 \neq Q_2$ we have

$$[A, Q_1] \cap [A, Q_2] = \varnothing$$

A grammar is called *LF* if each of its variables is left factored. As a matter of fact, *LF* grammars are the same as $LL(1)$ ones. This can be shown easily by the observation that $[A, Q]$ is nothing else but the union of the sets $\text{FIRST}_1(QW)$ for the appropriate words W, and thus the *LF* condition can be written as

$$\text{FIRST}_1(Q_1W) \cap \text{FIRST}_1(Q_2W) = \varnothing$$

for all W such that

$$S \overset{*}{\Rightarrow} XAW$$

for some $X \in V_T^*$. But this is the same as the condition of Theorem 9.3 for $k = 1$.

Another interesting subclass of context-free grammars is called $LR(k)$ grammars for *left to right scan, producing rightmost derivations with k symbol lookahead*. These grammars allow for *bottom-up* parsing deterministically with k symbol lookahead. Such a parsing process is represented by the arrangement of Figure 9.7. If the final portion of the contents of the push-down stack is equal to the right-hand side of some rule $A \rightarrow Q$, then the string Q may be replaced by the symbol A in the stack. (A becomes the top of the stack.) But it is not always practical to perform such a replacement because the input string may be derived in some other way. (For instance, we may have another rule $B \rightarrow QR$ which must be used for the actual input.) If the parser can always make a correct decision using the extra information on the contents of the stack and on the first k symbols of the remaining input, then the grammar is called $LR(k)$. The rule selected for application by an $LR(k)$ parser is called the *handle-production* and its right-hand side is the *handle* for the corresponding reduction.

The parser decides in each step whether a handle occurs on top of the stack, and, if so, it determines the production to be used for the reduction. Otherwise the next input symbol must be shifted from the input to the top of the stack. *This way the parser produces a rightmost derivation, though in reverse order.* Namely, if $A \rightarrow Q$ and $B \rightarrow R$ are used immediately one after the other for making reductions then we have some $X_1, X_2 \in (V_N \cup V_T)^*$ and $Y_1, Y_2 \in V_T^*$ with $|Y_2| \leqslant |Y_1|$ such that

$$X_2 B Y_2 \Rightarrow X_2 R Y_2 = X_1 A Y_1 \Rightarrow X_1 Q Y_1$$

Hence, in each step of this derivation the rightmost variable is replaced.

Definition 9.2 A context-free grammar $G = (V_N, V_T, S, F)$ is $LR(k)$ iff it does not contain the rule $S \rightarrow S$ and for every pair of rightmost derivations of the form

$$S \overset{*}{\Rightarrow} XAY \overset{*}{\Rightarrow} XQY$$

and

$$S \overset{*}{\Rightarrow} UBZ \overset{*}{\Rightarrow} URZ$$

where A and B are variables, X, U, Q, and R are in $(V_N \cup V_T)^*$, and Y and Z are in V_T^*, the equality

$$XQY: (|XQ| + k) = URZ: (|XQ| + k)$$

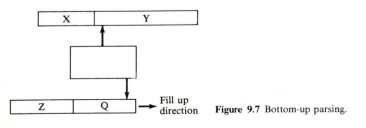

Figure 9.7 Bottom-up parsing.

implies

$$X = U, \quad A = B, \quad \text{and} \quad Q = R$$

Theorem 9.5 Every $LR(k)$ grammar is unambiguous.

PROOF Ambiguity means that there is some word P in $L(G)$ for which two different rightmost derivations exist. (Essentially different derivations remain different when they are rearranged into rightmost ones.) Let the last step of these derivations be

$$XAY \Rightarrow XQU = P$$

and

$$UBZ \Rightarrow URZ = P$$

respectively. By Definition 9.2 this implies $X = U$, $A = B$, and $Q = R$. So the last step is the same in both derivations. Continuing this reasoning we can see that the two derivations must be identical.

Example 9.7 Consider the context-free grammar

$$G = (\{S, A\}, \{a, +, *\}, S, F)$$

with the rules in F

$$S \rightarrow S + A \qquad A \rightarrow A * A$$

$$S \rightarrow A \qquad\quad A \rightarrow a$$

Suppose we want to parse the expression

$$a + a * a$$

from left to right and bottom up. Obviously, we apply the rule $A \rightarrow a$ first which gives

$$A + a * a \Rightarrow a + a * a$$

Now, before the application of $S \rightarrow A$ we have to examine the symbol next to A, because this rule should not be used if A is followed by $*$. Here we can apply $S \rightarrow A$ and $A \rightarrow a$ thereafter, so we get

$$S + A * a \Rightarrow S + a * a \Rightarrow A + a * a \Rightarrow a + a * a$$

where we can choose between $S \rightarrow A$ and $S \rightarrow S + A$. But consulting the next symbol $*$ reveals that neither of them should be applied because they would lead to a dead end. Instead, we have to read the input further and reach the sentential form

$$S + A * A$$

which in turn can be reduced to

$$S + A$$

and finally to

$$S$$

(None of the strings $S + S * a$ and $S * a$ can be reduced to S.) Using the same reasoning for arbitrary words in $L(G)$ we can show this grammar is $LR(1)$, but certainly not $LR(0)$. It is possible, however, to generate the same language by an $LR(0)$ grammar, though not very naturally.

Example 9.8 The following grammar

$$G = (\{S\}, \{a\}, S, \{S \rightarrow aSa, S \rightarrow a\})$$

is not $LR(k)$ for any k. Namely, for every k we have the rightmost derivations

$$S \overset{*}{\Rightarrow} a^k S a^k \Rightarrow a^{2k+1} \quad \text{and} \quad S \overset{*}{\Rightarrow} a^{k+1} S a^{k+1} \Rightarrow a^{2k+3}$$

where $|a^k a| + k = 2k + 1$ and thus,

$$a^{2k+1} : (2k + 1) = a^{2k+3} : (2k + 1)$$

but $a^k \neq a^{k+1}$ violates the $LR(k)$ condition of Definition 9.2. So we cannot always decide here which of the two rules is to be applied by looking ahead just a fixed number of symbols..

$LR(k)$ grammars are also applied in compiler design and have many interesting properties which we do not discuss here. We mention only that for every $LR(k)$ grammar we can find an equivalent deterministic pushdown automaton, and every deterministic pushdown automaton has an equivalent $LR(1)$ grammar. Therefore, only the cases $k = 0$ and $k = 1$ are really important. Also, it should be noted that there are much more efficient ways to check for the $LR(1)$ condition rather than use Definition 9.2 directly. Such methods are often used in compiler construction, and they can also be used to obtain the corresponding parsing algorithm for the given grammar at the same time provided that it is truly $LR(1)$. The usual model as shown in Figure 9.7 is, however, not a standard pushdown automaton in general because the entire string Q is replaced by a reduction step and not just the topmost symbol of the stack. Theoretically, this can be avoided by using additional states in the finite state control device together with λ-moves to implement one reduction step in several standard pushdown automaton moves. For practical purposes it is better to manipulate the stack more freely instead of using several states. For more details on parsing techniques we refer to the extensive literature.

TEN

DERIVATION LANGUAGES

Johnson M. Hart
György E. Révész

10.1 OPERATIONS ON DERIVATIONS

Derivations were introduced as a subject of study in Chapter 5, and the subject will now be examined in greater detail. A derivation in a type 0 grammar is regarded as a sequence of steps of the form

$$P = X_0 \Rightarrow X_1 \Rightarrow \cdots \Rightarrow X_n = Q \qquad (n \geqslant 0)$$

This derivation is said to be a derivation (in grammar G) from word P to word Q. Notice in the grammar $G = (V_N, V_T, S, F)$ that the initial symbol S and the distinction between nonterminal and terminal symbols are not really necessary. This leads to the definition.

> **Definition 10.1** A *rewriting system R* is a system $R = (V, F)$ where V is a finite alphabet and F is a finite set of ordered pairs (P, Q) such that $P \in V^+$ and $Q \in V^*$. If $G = (V_N, V_T, S, F)$ is a grammar, the *rewriting system* of G is $R_G = (V_N \cup V_T, F)$.

The Chomsky hierarchy of grammars (Definition 1.6) applies to rewriting systems.

As a convention, whenever $G = (V_N, V_T, S, F)$ is used, $V = V_N \cup V_T$ so that $R_G = (V, F)$. For any rewriting system $R = (V, F)$, V_F will have the usual meaning of being the names of the members of F. The names in V_F are often called *production names* or *production labels*.

Theorem 5.2 and its proof state that for every derivation $P \overset{*}{\Rightarrow} Q$ in a rewriting system, there is a leftmost (or canonical) derivation $P \overset{*}{\underset{\text{left}}{\Rightarrow}} Q$ with the same number of steps. Section 5.2 also states that every derivation is represented by a unique derivation graph, and each derivation graph represents a unique leftmost derivation. As a result of this, we can say that two derivations (represented as derivation sequences) are *equivalent* iff they have the same derivation graphs or leftmost derivations. Therefore, an entire equivalence class of derivations is represented by a derivation graph or a leftmost derivation (hence the term "canonical"). Two equivalent derivations differ only in the order of application of some productions.

When we refer to derivations in the future, it is often easiest to think not of a derivation sequence but of the graph or *canonical derivation* which represents the equivalence class of the derivation.

Two operations are fundamental for derivations, namely *composition* and *juxtaposition*. These concepts were originally formulated by Hotz [1966]. In order to develop these notions, it is first necessary to give names to derivations, and a rewriting system $R = (V, F)$ or a grammar $G = (V_N, V_T, S, F)$ with its rewriting system R_G (or simply R) is always assumed.

Definition 10.2 Let W_1 and W_2 be two derivations in rewriting system R such that

$$W_1: P_1 = X_0 \Rightarrow X_1 \Rightarrow \cdots \Rightarrow X_m = Q_1$$

and

$$W_2: P_2 = Y_0 \Rightarrow Y_1 \Rightarrow \cdots \Rightarrow Y_n = Q_2$$

The *juxtaposition* of W_1 with W_2, written as $W_1 \times W_2$, is the derivation

$$W_1 \times W_2: P_1 P_2 = X_0 Y_0 \Rightarrow X_1 Y_0 \Rightarrow \cdots \Rightarrow X_m Y_0 \Rightarrow X_m Y_1 \Rightarrow X_m Y_2$$
$$\Rightarrow \cdots \Rightarrow X_m Y_n = Q_1 Q_2$$

which is a derivation from $P_1 P_2$ to $Q_1 Q_2$.

Juxtaposition, then, is just the side-by-side application of the two derivations. If W_1 and W_2 are both leftmost, so is $W_1 \times W_2$. The derivation graph of $W_1 \times W_2$ is obtained by placing the graphs of W_1 and W_2 side by side.

Algebraically, juxtaposition is an associative but noncommutative operation. For every word $X \in V^*$, we associate the special derivation id_X, which is the derivation of length 0 (no steps) of X from itself, that is,

$$id_X: X \overset{*}{\Rightarrow} X$$

Thus id_λ is the identity operation for juxtaposition, so that if W is any derivation

$$id_\lambda \times W = W \times id_\lambda = W$$

Composition of derivations is the second algebraic operation and is now defined.

Definition 10.3 Let W_1 and W_2 be two derivations in rewriting system R such that

$$W_1: P = X_0 \Rightarrow X_1 \Rightarrow X_2 \Rightarrow \cdots \Rightarrow X_m = Q$$

and $\qquad W_2: Q = Y_0 \Rightarrow Y_1 \Rightarrow Y_2 \Rightarrow \cdots \Rightarrow Y_n = T$

The composition of W_2 with W_1, written $W_2 \circ W_1$, is the derivation

$$W_2 \circ W_1: P \overset{*}{\Rightarrow} Q \overset{*}{\Rightarrow} T$$

which is a derivation of T from P.

The composition of two derivations is obtained by applying one after the other. Composition can only be defined, of course, when the derived word of the first derivation is the start word of the second derivation. The derivation graph of the composition can be obtained by attaching the two graphs to each other in a way which will be made precise in the following sections.

Composition, like juxtaposition, is an associative, noncommutative operation on derivations. Note that if

$$W: P \Rightarrow Q$$

then $\qquad W \circ id_p = id_Q \circ W = W$

This is the first of two fundamental identities for derivations given by Hotz [1966]. The second identity is a little more difficult and will be proved formally in the next section. Let W_1, W_2, U_1, and U_2 be derivations. Then, if all the operations are defined,

$$(W_2 \times U_2) \circ (W_1 \times U_1) = (W_2 \circ W_1) \times (U_2 \circ U_1)$$

Hotz goes on to show that two derivations are equivalent iff one can be obtained from the other by a series of applications of these two identities.

The next two sections develop the concrete representation of equivalence classes, and the algebraic manipulation of these representations.

10.2 DERIVATION WORDS

For an algebraic treatment of derivation graphs (defined in Section 5.2) a rather compact string representation has been introduced by Hart [1975] by assigning to each derivation graph a unique word over $V \cup V_F$. Such words are called *derivation words*, and they have a number of interesting properties. Most important, the *composition* and *juxtaposition* of canonical derivations introduced by Hotz [1966] and defined in Section 10.1 can thus be defined explicitly as string manipulating operations on derivation words. Our presentation here and in the next section is based on the simplified definitions and related results developed by Révész [1977].

The set of all derivation words for a given rewriting system R is called the *derivation language* of R and will be denoted by $D(R)$. Two functions, *domain* and *codomain*, are fundamental for derivation languages. They can be defined independently of each other.

Definition 10.4 Let $R = (V, F)$ be a rewriting system. The *domain* of a word $W \in (V \cup V_F)^*$, $dom(W)$, is defined inductively as follows:

1) If $W \in V^*$, then $dom(W) = W$
2) If $dom(XY)$ is defined and $f: P \to Q$ is a rule in F, then $dom(XfQY) = dom(XY)$

The *codomain* of a word $W \in (V \cup V_F)^*$, $cod(W)$, is defined inductively as follows:

1) If $W \in V^*$, then $cod(W) = W$
2) If $cod(XY)$ is defined and $f: P \to Q$ is a rule in F, then $cod(XPfY) = cod(XY)$

Note that $dom(W)$ and $cod(W)$ are in V^* whenever they are defined. It can be shown by induction on the number of occurrences of production names in W that these definitions are precise. For, if W has two different evaluations with $X_1f_1Q_1Y_1$ and $X_2f_2Q_2Y_2$, then the two parts f_1Q_1 and f_2Q_2 cannot overlap; that is, either $W = Xf_1Q_1Zf_2Q_2Y$ or $W = Xf_2Q_2Zf_1Q_1Y$ for some X, Z, Y in $(V \cup V_F)^*$. Therefore,

$$dom(W) = dom(X_1Y_1) = dom(X_2Y_2) = dom(XZY)$$

which is precise by the induction hypothesis. The argument is similar for $cod(W)$.

These two functions have the self-embedding property expressed by the next theorem.

Theorem 10.1 If W_1, W_2, and W_3 are in $(V \cup V_F)^*$ then
$$dom(W_1W_2W_3) = dom(W_1dom(W_2)W_3)$$
and
$$cod(W_1W_2W_3) = cod(W_1cod(W_2)W_3)$$

whenever the right-hand side of the corresponding equation is defined.

PROOF We use induction on the number of occurrences of production names in W_2.

Basis: If W_2 is in V^* then the assertion is trivial.

Induction step: Suppose that W_2 contains n production names ($n \geqslant 1$) and $dom(W_1 dom(W_2) W_3)$ is defined. Then $dom(W_2)$ also must be defined, hence $W_2 = XfQY$ for some $X, Y \in (V \cup V_F)^*$ and $f: P \rightarrow Q \in F$ with $dom(W_2) = dom(XfQY) = dom(XY)$. Then we have

$$dom(W_1 W_2 W_3) = dom(W_1 XfQY W_3) = dom(W_1 XY W_3)$$

by definition and

$$dom(W_1 XY W_3) = dom(W_1 dom(XY) W_3)$$

by the induction hypothesis which completes the proof.

The second equation can be shown by a similar argument.

Definition 10.5 Let $R = (V, F)$ be a rewriting system. The derivation language of R, $D(R)$, is defined inductively as follows:

1) If $W \in V^*$, then $W \in D(R)$
2) If $XY \in D(R)$ with $cod(X)$ and $cod(Y)$ being defined and $f: P \rightarrow Q \in F$ such that $cod(X) = ZP$ for some $Z \in V^*$, then $XfQY \in D(R)$
3) Nothing else belongs to $D(R)$

The elements of $D(R)$ are called derivation words of R.

Consider for example the derivation graph in Figure 5.1 (Section 5.2). This graph will be represented by the word

$$Sf_1 Sf_1 SAbAbf_2 af_3 ABBa$$

In order to show the domain and codomain of this derivation word, we parenthesize its components in both ways.

Domain: $S(f_1 S(f_1 SAb) Ab)(f_2 a(f_3 AB) Ba)$

Codomain: $(Sf_1)(Sf_1) S(A(bAbf_2) af_3) ABBa$

As can be seen in this example the domain and the codomain correspond to the starting and the ending nodes, respectively. This is true for every derivation word as is shown below.

Theorem 10.2 If $W \in D(R)$, then $dom(W) \underset{R}{\overset{*}{\Rightarrow}} cod(W)$.

PROOF This is shown by induction on the number of occurrences of production names in W. For $W \in V^*$ the assertion is trivial. Let W

contain n production names ($n \geqslant 1$). Then $W = XfQY$ for some $X, Y \in (V \cup V_F)^*$; $f \in V_F$, and $Q \in V^*$ as required by the definition of $D(R)$. Hence,

$$dom(W) = dom(XfQY) = dom(XY)$$

and $$dom(XY) \underset{R}{\overset{*}{\Rightarrow}} cod(XY)$$

by the induction hypothesis. Further, by Theorem 10.1 we have

$$
\begin{aligned}
cod(XY) &= cod(cod(X)Y) = cod(cod(X)cod(Y)) \\
&= cod(X)cod(Y) = ZPcod(Y) \underset{R}{\Rightarrow} ZQcod(Y) \\
&= cod(ZQcod(Y)) = cod(ZQY) \\
&= cod(Zcod(Pf)QY) = cod(ZPfQY) \\
&= cod(cod(ZP)fQY) = cod(XfQY) = cod(W)
\end{aligned}
$$

which completes the proof.

Corollary The language generated by a phrase-structure grammar $G = (V_N, V_T, S, F)$ can be given as

$$L(G) = \{cod(W) | W \in D(R_G), dom(W) = S, cod(W) \in V_T^*\}$$

REMARK Theorem 10.2 implies also the fact that for every derivation word $W \in D(R)$, both $dom(W)$ and $cod(W)$ are defined. For an arbitrary word in $(V \cup V_F)^*$, neither of them has to be defined, but even if one of them is defined, the other may still be undefined. If, for instance, $f: P \to Q \in F$ then obviously $dom(fQ) = \lambda$, but $cod(fQ)$ is undefined. (Similarly, $cod(Pf) = \lambda$, but $dom(Pf)$ is undefined except when $Q = \lambda$.) On the other hand, it can be shown that if both $dom(W)$ and $cod(W)$ are defined, then W must be in $D(R)$ which gives the following characterization theorem.

Theorem 10.3 A word $W \in (V \cup V_F)^*$ belongs to $D(R)$ iff both $dom(W)$ and $cod(W)$ are defined.

PROOF In view of the previous remark it is enough to show the if part. For this we use induction on the number of occurrences of production names in W.

Basis: For $W \in V^*$ the assertion is trivial.
Induction Step: Let W contain n production names ($n \geqslant 1$) and both $dom(W)$ and $cod(W)$ be defined. Then there must occur some f in W such that $W = XfQY$ and $dom(W) = dom(XY)$. If there is more than

one such f then we choose the rightmost one. (Namely, the computation of $dom(W)$ is performed by successive cancellation of production names occurring in W together with a possibly empty substring from the right context of each production name. Hence, the rightmost f in W may be canceled first.)

For this particular f we have $cod(W) = cod(XfQY) = cod(cod(Xf)QY)$, provided that $cod(Xf)$ is defined. But this must be the case, otherwise in the course of the computation of $cod(W)$ this f could not be canceled at all. Moreover, any truncation of W from the right preserves the computability (if not the value) of the codomain function. (Symmetrically, left truncation preserves the computability of the domain function.) Therefore, $cod(X)$ must be also defined and $cod(X) = ZP$ for some $Z, P \in V^*$ with $f: P \to Q \in F$. This means that $cod(XY)$ is also defined as $cod(XY) = cod(cod(X) cod(Y)) = cod(X)cod(Y)$.

Now, $XY \in D(R)$ by the induction hypothesis and thus, $W = XfQY \in D(R)$ by the definition of $D(R)$.

To conclude this section we want to note that the term "*derivation language*" is used by some authors in a different sense. In particular, Penttonen [1974b] uses this term synonymously with Szilard languages which are the following:

Let $G = (V_N, V_T, S, F)$ be a phrase-structure grammar with a unique name in V_F assigned to each production in F. The Szilard language associated with G is defined by

$$Sz(G) = \{ f_1 \cdots f_n | f_1, \dots, f_n \text{ is a terminal derivation in } G \}$$

Hence the Szilard language of G depends on S and V_T, and $Sz(G) \subset V_F^*$. Another important difference is that the terminal derivation in the definition of $Sz(G)$ need not be leftmost; hence, there can be several words of $Sz(G)$ corresponding to the same canonical derivation. (This aspect of the Szilard language can be used in dealing with firing sequences in Petri nets.)

Szilard languages are also related to the idea of *control sets* defined by Ginsburg and Spanier [1968], where a subset of V_F^* is defined on its own as being the control set of permissible derivations in G.

10.3 ALGEBRAIC PROPERTIES OF THE FUNDAMENTAL OPERATIONS

Two fundamental operations, *juxtaposition* and *composition*, have been defined on derivations in Section 10.1. Now, we have to define them as operations on derivation words. Incidentally, we shall define them on words in $(V \cup V_F)^*$

not necessarily belonging to $D(R)$. For juxtaposition this can be done without any restriction but for composition, which is more delicate, the components should be derivation words or very close to such words. This slightly generalized version of the composition leads to a more elegant recursive definition.

Definition 10.6 Let W_1 and W_2 be in $(V \cup V_F)^*$. Then the *juxtaposition* of W_1 with W_2, written $W_1 \times W_2$, is the catenation of W_1 and W_2.

Theorem 10.4 If W_1 and W_2 are in $D(R)$ then $W_1 \times W_2$ is also in $D(R)$.

PROOF By Theorem 10.3 $dom(W_1)$, $dom(W_2)$, $cod(W_1)$, and $cod(W_2)$ are all defined. By the above definition $W_1 \times W_2$ is simply the word $W_1 W_2 \in (V \cup V_F)^*$. Therefore, using Theorem 10.1 we get

$$dom(W_1 W_2) = dom(dom(W_1)\,dom(W_2)) = dom(W_1)\,dom(W_2)$$

and $\quad cod(W_1 W_2) = cod(cod(W_1)\,cod(W_2)) = cod(W_1)\,cod(W_2)$

Hence by Theorem 10.3, $W_1 W_2$ is in $D(R)$, and this completes the proof.

The composition of derivations can be illustrated by the following example. Let $G = (V_N, V_T, S, F)$ be a phrase structure grammar with $V_N = \{S, A, B\}$, $V_T = \{a, b\}$ and with the following rules:

$$f_1: S \to ABB \qquad\qquad f_2: S \to BA$$
$$f_3: AB \to SAB \qquad\qquad f_4: BAS \to AA$$
$$f_5: A \to a \qquad\qquad f_6: B \to b$$

Then the two derivations given in Figure 10.1 can be composed into one as shown in Figure 10.2.

In general we can define the composition of two derivations if the ending nodes of the first graph, in the order from left to right, are the same as the starting nodes of the second graph, again taking into account their order from left to right. (Extra nodes on both sides can be added, if necessary, with the assumption that they are starting and ending nodes at the same time.) The question is how to obtain the derivation word of the composed graph from the derivation words of the two component graphs. The two components in our example are represented by

$$ABf_3 SABf_3 Sf_2 BAAB$$

and $\qquad Sf_2 BABf_3 Sf_2 BAABf_3 Sf_4 AAAB$

and their composition corresponds to

$$ABf_3 Sf_2 BAABf_3 Sf_2 Bf_3 Sf_2 BAABf_3 Sf_4 AAABAAB$$

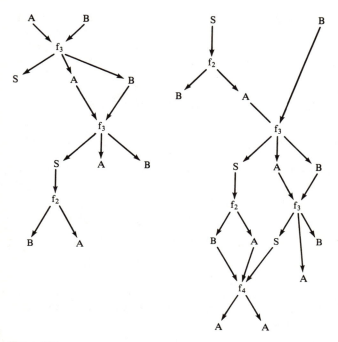

Figure 10.1

as can be figured out from the given graphs. A precise definition for the general case can be given as follows.

Definition 10.7 Let W_1 and W_2 be in $(V \cup V_F)^*$ such that $cod(W_1)$ and $dom(W_2)$ are defined, $cod(W_1) = dom(W_2)$, and $cod(W_1) \neq \lambda$ if $W_1 \neq \lambda$. Then the composition of W_2 with W_1, written $W_2 \circ W_1$, is defined recursively as follows:

1) If $W_1 = \lambda$, then $W_2 \circ W_1 = W_2$
2) If there is some $X \in V^*$ and there are some $Y_1, Y_2, Z_1, Z_1 \in (V \cup V_F)^*$ such that $W_1 = Y_1 X Z_1$ with $cod(W_1) = X cod(Z_1)$ and

$$W_2 = Y_2 X Z_2 \quad \text{with } dom(Y_2) = \lambda$$

then $W_2 \circ W_1 = Y_2 Y_1 X (Z_2 \circ Z_1)$

Note that $cod(Y_1) = \lambda$ always holds here, because

$$cod(W_1) = cod(Y_1 X Z_1) = cod(Y_1 X cod(Z_1))$$
$$= cod(Y_1 cod(W_1)) = cod(Y_1) cod(W_1)$$

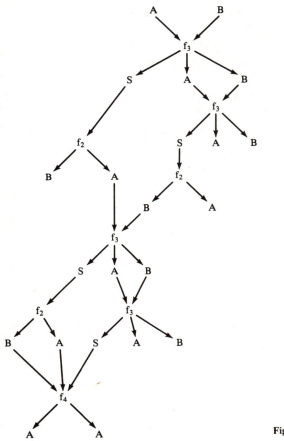

Figure 10.2

provided that $cod(Y_1)$ is defined. But this must be the case since $cod(W_1)$ is defined and any truncation of W_1 from the right preserves the computability (if not the value) of the codomain function.

On the other hand, W_1 and W_2 need not be derivation words. If, for instance, $f: AB \rightarrow CD \in F$ and U_1 and U_2 are derivation words with $cod(U_1) = dom(U_2)$ then $fCDU_2 \circ U_1$ is also defined though $fCDU_2$ is not a derivation word since $cod(fCDU_2)$ is undefined.

A fundamental property of the composition is expressed by the following theorem.

Theorem 10.5: Composition Theorem If W_1 and W_2 are in $D(R)$ and their composition $W_2 \circ W_1$ is defined, then

$$dom(W_2 \circ W_1) = dom(W_1)$$

and

$$cod(W_2 \circ W_1) = cod(W_2)$$

PROOF For $W_1 = \lambda$ the assertion is trivial. Otherwise recursion scheme 2 must be applied a finite number of times, which yields the decompositions

$$W_1 = Y_{1,1}X_1 \cdots Y_{1,k}X_kZ_{1,k}$$
$$W_2 = Y_{2,1}X_1 \cdots Y_{2,k}X_kZ_{2,k}$$

and

$$W_2 \circ W_1 = Y_{2,1}Y_{1,1}X_1 \cdots Y_{2,k}Y_{1,k}X(Z_{2,k} \circ Z_{1,k})$$

where $cod(Y_{1,i}) = dom(Y_{2,i}) = \lambda$ for $i = 1, \ldots, k$, and $dom(Z_{2,k}) = Z_{1,k} = \lambda$. Hence, using Theorem 10.1, we get

$$
\begin{aligned}
dom(W_2 \circ W_1) &= dom(Y_{2,1}Y_{1,1}X_1 \cdots Y_{2,k}Y_{1,k}X_k(Z_{2,k} \circ Z_{1,k})) \\
&= dom(dom(Y_{2,1})Y_{1,1}X_1 \cdots dom(Y_{2,k})Y_{1,k}X_k dom(Z_{2,k})) \\
&= dom(Y_{1,1}X_1 \cdots Y_{1,k}X_k) = dom(W_1)
\end{aligned}
$$

and similarly

$$
\begin{aligned}
cod(W_2 \circ W_1) &= cod(Y_{2,1}cod(Y_{1,1})X_1 \cdots Y_{2,k}cod(Y_{1,k})X_kZ_{2,k}) \\
&= cod(Y_{2,1}X_1 \cdots Y_{2,k}X_kZ_{2,k}) = cod(W_2)
\end{aligned}
$$

which completes the proof.

Theorem 10.6: Juxtaposition Theorem For every $W_1, W_2, U_1, U_2 \in (V \cup V_F)^*$ the identity

$$(W_2 \times U_2) \circ (W_1 \times U_1) = (W_2 \circ W_1) \times (U_2 \circ U_1)$$

holds whenever its right-hand side is defined.

PROOF If $W_1 = \lambda$ then the assertion is trivial. Otherwise $W_1 = Y_1XZ_1$ and $W_2 = Y_2XZ_2$ for some Y_1, Y_2, X, Z_1, and Z_2 according to condition 2 of Definition 10.7. Hence

$$
\begin{aligned}
(W_2U_2) \circ (W_1U_1) &= Y_2Y_1X((Z_2U_2) \circ (Z_1U_1)) \\
&= Y_2Y_1X(Z_1 \circ Z_2)(U_2 \circ U_1) = (W_2 \circ W_1)(U_2 \circ U_1)
\end{aligned}
$$

by definition and by the induction hypothesis.

Clearly, the operations of composition and juxtaposition can be extended to more than two operands, and analogous theorems will hold for this case.

Example 10.1 Let us consider the rewriting system containing the following rules:

$$f_1: AB \rightarrow CAB$$
$$f_2: C \rightarrow BA$$
$$f_3: BAA \rightarrow CBA$$

Then
$$W_1 = ABf_1CABf_1Cf_2BAAB$$
and
$$W_2 = Cf_2BABf_1CABAAf_3CBAB$$

are derivation words with

$$cod(W_1) = dom(W_2) = CBAAB$$

To find the composition $W_2 \circ W_1$ we shall use the decompositions

$$W_1 = Y_1XZ_1 \text{ with } Y_1 = ABf_1, X = C, Z_1 = ABf_1Cf_2BAAB$$
and $\quad W_2 = Y_2XZ_2 \text{ with } Y_2 = \lambda, X = C, Z_2 = f_2BABf_1CABAAf_3CBAB$

It can be observed that Z_2 is not a derivation word, but $Z_2 \circ Z_1$ is still defined. The repeated application of recursion scheme 2 will yield

$$W_2 \circ W_1 = ABf_1Cf_2BAABf_1Cf_2Bf_1CABAAf_3CBAB$$

which is again a derivation word with

$$dom(W_2 \circ W_1) = dom(W_1) \quad \text{and} \quad cod(W_2 \circ W_1) = cod(W_2)$$

This means, of course, that

$$AB = dom(W_1) \overset{*}{\underset{R}{\Rightarrow}} cod(W_2) = BCACBAB$$

REMARK Our results so far should make it clear that our derivation words have some very nice properties similar to those of the embedded list (parenthesized string) representation of derivation trees. In particular, the computation of the domain and codomain functions via their self-embedding property is analogous with the tree computation.

10.4 CANONICAL DERIVATIONS AND GRAPH TRAVERSALS

Theorem 10.6 at the end of the last section shows that derivation words remain unchanged under change of order of production rule application in a derivation sequence. Likewise, such changes in order do not affect the derivation graph, so we would expect that both the derivation word and the derivation graph are the same for every derivation sequence in an equivalence class. Every equivalence class of derivations has a unique canonical derivation (Theorem

5.2), and we now show how the canonical derivation can be obtained directly from either a derivation word or a graph. In the process, we develop a graph traversal technique that is a generalization of preorder tree traversal (a well-known data structure algorithm) and show that derivation words generalize the familiar prefix representation of context-free derivation trees.

Definition 10.8 Let $R = (V, F)$ be a rewriting system with $W \in (V \cup V_F)^*$. If W can be written as $W = X_1 f Q X_2$ where $f: P \to Q \in F$, $X_1 \in (V \cup V_F)^*$, and $X_2 \in V^*$, we say that W *right reduces* to $X_1 X_2$, written $W = X_1 f Q X_2 \vdash X_1 X_2$.

Right reduction corresponds to removal of the rightmost production in the word, and by Definition 10.4, $dom(W) = dom(X_1 X_2)$. The right reduction of a word, if it exists, is unique.

The canonical or leftmost derivation sequence can now be obtained from the sequence of right reductions of a derivation word.

Theorem 10.7 Let $R = (V, F)$ be a rewriting system with $W \in D(R)$. If
$$W = W_n \vdash W_{n-1} \vdash \cdots W_1 = dom(W)$$
then
$$cod(W_1) \Rightarrow cod(W_2) \Rightarrow \cdots \Rightarrow cod(W_n)$$

is a canonical derivation sequence. The production rule applied at each step is the one involved in the right reduction.

PROOF First, note that such a sequence of reductions always exists for a derivation word by the definitions of the previous sections.
Assume that
$$W_{i+2} = U_{i+1} f_{i+1} Q_{i+1} Y_{i+1} \vdash U_{i+1} Y_{i+1} = W_{i+1} = U_i f_i Q_i Y_i \vdash U_i Y_i = W_i$$
where, of course, $Y_{i+1}, Y_i \in V^*$ and $f_{i+1}: P_{i+1} \to Q_{i+1}, f_i: P_i \to Q_i \in F$. In addition, write
$$cod(W_i) = X_i P_i Y_i$$
and
$$cod(W_{i+1}) = X_{i+1} P_{i+1} Y_{i+1} = X_i Q_i Y_i$$
It is now necessary to show (Definition 5.1) that $|X_i| < |X_{i+1} P_{i+1}|$. By the construction above, it must be possible to write
$$W_{i+2} = U_{i+1} f_{i+1} Q_{i+1} Y_{i+1} = U'_{i+1} f_i U''_{i+1} f_{i+1} Q_{i+1} Y_{i+1}$$
with $U''_{i+1} \in (V_N \cup V_T)^+$ (assuming no nontrivial null righthand sides using Theorem 5.1) and $U''_{i+1} Y_{i+1} = Q_i Y_i$. Hence $|Q_i Y_i| > |Y_{i+1}|$. Since

$$X_i Q_i Y_i = X_{i+1} P_{i+1} Y_{i+1},$$
$$|X_i| + |Q_i| + |Y_i| = |X_{i+1}| + |P_{i+1}| + |Y_{i+1}|$$
and
$$|X_i| < |X_{i+1} P_{i+1}|$$

completing the proof.

Theorem 10.7 shows, among other things, that the production names occur in a derivation word (in left-to-right order) in exactly the same order as they are used in the canonical derivation. The canonical derivation can be found very quickly from a derivation word (the time is a linear function of the number of productions), and the derivation word can be computed directly from any derivation (Definition 10.5), as can the derivation graph. The next problem is to show how to construct the derivation word by traversing the graph in some manner and reading off the labels. The technique for doing this is developed next.

Given a derivation graph, consider any two nodes, a and b, of the directed graph (Figures 5.1, 10.1, and 10.2 are good references for this).

There are three distinct possibilities for nodes a and b, assuming they are distinct.

1) a is *below* b, in the sense that there is a downward directed path from b to a
2) b is below a
3) There is no path between a and b, and one of the two nodes is to the "left" of the other

These three possibilities are mutually exclusive. We express the fact that node a is below node b by use of the relation B and write $(a, b) \in B$. If a is to the left of b, then write $(a, b) \in L$, where L is a relation.

As a derivation graph is constructed inductively from a derivation sequence, the B and L relations can be constructed as well. Both relations are intended to be *irreflexive* (i.e., no node is related to itself) but are *transitive* (e.g., if $(a, b) \in B$ and $(b, c) \in B$, then $(a, c) \in B$). The specific general properties of these relations are expressed by the following definition.

Definition 10.9 Let N be a finite set of nodes and B and L be two relations on N such that

1) B and L are transitive
2) $\langle B, L, B', L', eq \rangle$ is a partition of $N \times N$

$\quad$ ($B' = \{(a, b) | (b, a) \in B\}$ and eq is the equality relation.)
$\quad$ The system $D = (N, B, L)$ is said to be a *doubly ordered graph* (DOG) over N.

Recall that a partition of a set is a division of the set into disjoint subsets. The partition property assures, among other things, that the B and L relations are irreflexive.

By a rather tedious but straightforward inductive proof, we can establish the following:

Fact: The B and L relations computed for a derivation graph form a DOG.

The DOG properties are all that we need to perform the traversals of derivation graphs. Intuitively, a derivation graph is a labeled DOG (labels chosen from $V \cup V_F$). It is necessary to traverse the graph in such a way that no node is visited until all of its ancestors with respect to both the B^r and L relations have been visited.

Theorem 10.8 Let $D = (N, B, L)$ be a doubly ordered graph. N can be written uniquely as

$$N = \{n_1, n_2, \ldots, n_k\}$$

such that $i > j$ holds iff

$$(n_i, n_j) \in B \quad \text{or} \quad (n_i, n_j) \in L^r \left(\text{that is, } (n_i, n_j) \in B \cup L^r\right)$$

PROOF It is necessary to show that $B \cup L^r$ is a total order on N; i.e., if $n, m \in N$ with $n \neq m$, then either $(m, n) \in B \cup L^r$ or $(n, m) \in B \cup L^r$ (most definitions of "total order" require that a total order be reflexive, but this is a minor detail.) This fact follows immediately from the partition of $N \times N$. In addition, we must show that $B \cup L^r$ is transitive. To do this, assume that $m, n, p \in N$ with

$$(m, n) \in B \cup L^r \quad \text{and} \quad (n, p) \in B \cup L^r$$

and show that $(m, p) \in B \cup L^r$. If (m, n) and (n, p) are both in B (or both in L^r) the proof is immediate by the transitivity of B and L. The problem occurs if $(m, n) \in B$ and $(n, p) \in L^r$ (or vice versa). Clearly $m \neq p$, so if $(m, p) \notin B \cup L^r$ there are only two other possibilities, both leading to a contradiction:

1) $(m, p) \in B^r$, that is, $(p, m) \in B$, so that $(p, n) \in B$
2) $(m, p) \in L$; that is, $(p, m) \in L'$, so that $(n, m) \in L^r$

Theorem 10.8 allows us to find a derivation word by traversing the graph in such a way that no node is visited until all of the nodes it is below are visited. At any point, the leftmost of all the eligible nodes is selected to be visited. If the labels of the nodes are listed in the order of traversal, the derivation word results. This is summarized in the next theorem.

Theorem 10.9 Let $R = (V, F)$ be a rewriting system with W being a derivation graph in R and $D = (N, B, L)$ the DOG of the derivation graph of W. In addition, let $l: N \rightarrow V \cup V_F$ be the labeling function of W. If N is written as $N = \langle n_1, \ldots, n_k \rangle$ such that $(n_i, n_{i+1}) \in B \cup L^r$ $(1 \leqslant i < k)$ and l is regarded as a homomorphism, then

$$l(n_1 n_2 \cdots n_k)$$

is the derivation word corresponding to W.

PROOF The proof is by induction on the number of productions using the inductive definitions of a derivation word (Definition 10.5). In each step of the construction, it is necessary to define the resulting B and L relations in terms of the existing one. The complete proof is left as an exercise at the end of this section.

Theorem 10.9 establishes the concept of one production being "dependent" upon another one (in terms of the B relation) and of one production being "independent of" and "left of" another (using the L relation). Theorem 10.7 then shows that these relations capture exactly the properties required for the canonical (leftmost) derivation.

EXERCISES

10.1 Give the necessary and sufficient conditions on the B and L relations so that a DOG forms a tree.

10.2 Complete the proof that a derivation graph does indeed form a DOG.

10.3 Complete the proof of Theorem 10.9.

10.4 Given any derivation graph representing a derivation, give a suitable definition of a "subderivation" of this derivation as any partially complete derivation where any order of rule application (not necessarily leftmost) has been used. Show that the set of subderivations of a derivation form a lattice under suitable operations. *Hint*: use the DOG properties of the graph and use union and intersection (properly applied to graphs) as the lattice operations.

10.5 Give a suitable definition of rightmost (or right canonical) derivations. How could you obtain a rightmost derivation from a derivation graph or word?

10.5 THE CONTEXT-SENSITIVITY OF DERIVATION LANGUAGES

The derivation language, $D(R)$, of a rewriting system is defined inductively in Definition 10.5. The subsequent ease with which derivation words were processed (to perform the algebraic operations and to form graphs, for instance) suggests that the problem of recognizing a member of $D(R)$ should be easy. It will be demonstrated in this section that derivation languages are not only context-sensitive, but they can be recognized by a deterministic linear bounded automaton, as well. (See Section 6.3 for a discussion of linear bounded automata.)

Theorem 10.10 Let $R = (V, F)$ be a rewriting system. Then there is a deterministic linear bounded automaton (lba for short), A, with input alphabet $V \cup V_F$ such that $D(R) = L(A)$.

PROOF The basis for the construction of the lba is Theorem 10.3 which states that $W \in D(R)$ iff both $dom(W)$ and $cod(W)$ are defined. The lba is constructed so that it uses the two pushdown stacks in two separate steps:

1) Compute $dom(W)$ if it exists. This is done in the manner suggested by Definition 10.8. The input word W is scanned from the right to find the first production rule at the right end, and W is right reduced so that

$$W = X_1 f Q X_2 \vdash X_1 X_2$$

with X_1 on the first stack and fQX_2 on the second, fQ is easily erased (with a check to verify that there is some rule $f: P \to Q \in F$). The process is repeated on $X_1 X_2$ until there are no production rules left and $dom(W)$ (if it exists) is the concatenated contents of the two stacks. Theorem 10.7 assures the validity of this.
2) Compute $cod(W)$ if it exists. An entirely symmetrical operation, starting from the left end of W, is used.

Corollary For any phrase structure grammar, G, $D(G)$ is a deterministic context-sensitive language.

A context-sensitive grammar can be constructed directly for $D(R)$. The form of the construction is from Hart [1976]. The grammar given here will just generate those $W \in D(R)$ with $dom(W) = S$ (the initial symbol). Therefore, we start with a phrase structure grammar instead of a rewriting system.

The reader can easily generalize the construction to give all of $D(R)$. We assume that some λ-free phrase structure grammar $G = (V_N, V_T, S, F)$ is given. A new (length increasing) grammar, $G_D = (V_N^D, V_T^D, S'', F^D)$ is now constructed. The following notation is used.

1) $V = V_N \cup V_T$
2) $V' = \{v' \mid v \in V\}$
3) $V'' = \{v'' \mid v \in V\}$ (note that S'' is the start symbol)
4) If F is written as $F = \{f_1, \ldots, f_k\}$, then $W = \{v^{(i)} \mid v \in V, f_i \in F\}$

G_D then consists of the following components:

1) $V_T^D = V_N \cup V_T \cup V_F$
2) $V_N^D = V' \cup V'' \cup W$
3) F_D consists of the following productions:
   a) $v' \to v$ and $v'' \to v$ for all $v \in V$
   b) for each production $f_i \in F$ of the form

$$a_1 a_2 \cdots a_m \to b_1 \cdots b_n \ (a_i, b_j \in V; 1 \leqslant i \leqslant m, 1 \leqslant j \leqslant n)$$

add the following productions to F_D

$$a_1'' \to a_1' a_1^{(i)}$$

$$a_j^{(i)} x \to x a_j^{(i)} \quad \text{for all } x \in V' \cup F \quad \text{and} \quad 1 \leqslant j \leqslant m - 1$$

$$a_j^{(i)} a_{j+1}'' \to a_{j+1}' a_{j+1}^{(i)} \text{ for } 1 \leqslant j \leqslant m - 1$$

$$a_m^{(i)} \to f_i b_1'' b_2'' \cdots b_n''$$

EXERCISES

10.6 Show that $D(G)$ is context-free iff G is context-free.

10.7 Modify G_D above so that it generates

$$\{ W \in D(G) | dom(W) = S \quad \text{and} \quad cod(W) \in V_T^* \}$$

Thus, only derivation words of derivations of $L(G)$ are generated.

10.6 DERIVATIONS IN CONTEXT-SENSITIVE GRAMMARS

Context-sensitive, or type 1, grammars were defined in Section 1.2 so that all rewriting rules are of the form

$$Q_1 A Q_2 \to Q_1 P Q_2$$

with $A \in V_N$ and $Q_1, Q_2, P \in (V_N \cup V_T)^*$ (except that P can be λ only if the rule is $S \to \lambda$). Theorem 4.1 showed that every length increasing grammar is equivalent to a context-sensitive grammar in the sense that they generate the same language. Such equivalence is called *weak equivalence* for there is no implied correspondence between derivation graphs.

There is, in fact, a question as to exactly what is meant by a derivation in a context-sensitive grammar. The main question is whether or not the context symbols are rewritten or are merely used to permit the application of a production rule. Consider, for instance, the simple context-sensitive grammar with three rules:

$$f_1: S \to aAbBc$$

$$f_2: aAb \to axb$$

$$f_3: bBc \to byc$$

If the grammar is regarded as a length increasing type 0 grammar, then there are two distinct derivation graphs (hence canonical derivations or derivation words) for the word

$$axbyc$$

as the reader can quickly verify. This would imply that even this simple grammar is ambiguous.

On the other hand, one could take the view that a context-sensitive rewriting rule such as

$$Q_1 A Q_2 \to Q_1 P Q_2$$

is a context-free production rule

$$A \to P$$

which can only be used when A appears in the left context, Q_1, and right context, Q_2. To emphasize this interpretation, context-sensitive rewrite rules will now be put in the form:

$$A \to P | Q_1 {}_- Q_2$$

With this interpretation of the meaning of derivations in context-sensitive grammars, it is now necessary to develop the correct form of derivation graphs and to see if there is a meaningful concept of *canonical derivation*. The solution to these questions was given by Hart [1980] and will now be described.

If $G = (V_N, V_T, F, S)$ is a context-sensitive grammar, then a derivation in G is of the form

$$W_0 \Rightarrow W_1 \Rightarrow \cdots \Rightarrow W_n$$

where, for $i = 0, 1, \ldots, n - 1$, there are words $A_i \in V_N$ and $Q_i, R_i, P_i, X_i, Y_i \in (V_N \cup V_T)^*$ such that

$$A_i \to P_i | Q_i {}_- R_i \in F$$

$$W_i = X_i Q_i A_i R_i Y_i$$

$$W_{i+1} = X_i Q_i P_i R_i Y_i$$

This statement just follows from the definition. The conditions for a canonical, or leftmost, derivation are now stated.

Definition 10.10 Let $G = (V_N, V_T, F, S)$ be a CSG. A derivation in G,

$$W_0 \Rightarrow W_1 \Rightarrow \cdots \Rightarrow W_n$$

is said to be context-sensitive leftmost (or canonical) if $W_i = X_i Q_i A_i R_i Y_i$, $W_{i+1} = X_i Q_i P_i R_i Y_i$ with $A_i \rightarrow P_i | Q_i {}_- R_i \in F$ (for $i = 0, 1, \ldots, n - 1$) and, for each i, we have either

$$|X_{i+1} Q_{i+1} A_{i+1} R_{i+1}| > |X_i Q_i| \qquad \text{or} \qquad |X_{i+1} Q_{i+1}| \geqslant |X_i|$$

This definition should be compared carefully to Definition 5.1 (the definition of a type 0 leftmost derivation). Also, the reader should verify that if G is context-free (with null context strings) the definition becomes the standard one for leftmost context-free derivations.

To understand the meaning of this definition, suppose that for some i the canonical condition is violated, so that

$$|X_{i+1} Q_{i+1} A_{i+1} R_{i+1}| \leqslant |X_i Q_i|$$

and $\qquad |X_{i+1} Q_{i+1}| < |X_i| \qquad (\text{or } |X_{i+1} Q_{i+1} A_{i+1}| \leqslant |X_i|)$

But then, the rule f_{i+1}: $A_{i+1} \rightarrow P_{i+1} | Q_{i+1} {}_- R_{i+1}$ could have been applied before f_i: $A_i \rightarrow P_i | Q_i {}_- R_i$ since all of the context symbols for f_{i+1} appear to the left of A_i (the first inequality), and the rewritten symbol, A_{i+1}, does not appear in or alter the context Q_i to be used by f_i (the second inequality). Hence, f_{i+1} could be applied before f_i without interference to f_i.

Example 10.2 Consider the context-sensitive grammar with start symbol S and the following productions:

f_1: $S \rightarrow ABC$ (There are no context symbols, so they are not shown.)

f_2: $A \rightarrow XY$

f_3: $B \rightarrow DE | A {}_-$

f_4: $E \rightarrow AA | {}_- C$

f_5: $C \rightarrow FG$

One possible derivation sequence in this grammar is shown below with the rewritten symbol underlined and the rewrite rule placed over the $\Rightarrow$.

$$\underline{S} \overset{f_1}{\Rightarrow} A\underline{B}C \overset{f_3}{\Rightarrow} AD\underline{E}C \overset{f_4}{\Rightarrow} ADAA\underline{C} \overset{f_5}{\Rightarrow} \underline{A}DAAFG \overset{f_2}{\Rightarrow} XYDAAFG$$

This derivation applies the rules in the order

1) $f_1 f_3 f_4 f_5 f_2$

Two other *equivalent* sequences are

2) $f_1 f_3 f_2 f_4 f_5$
3) $f_1 f_3 f_4 f_2 f_5$

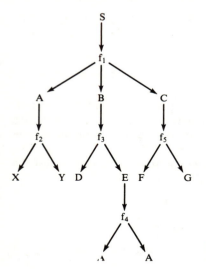

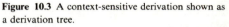

Figure 10.3 A context-sensitive derivation shown as a derivation tree.

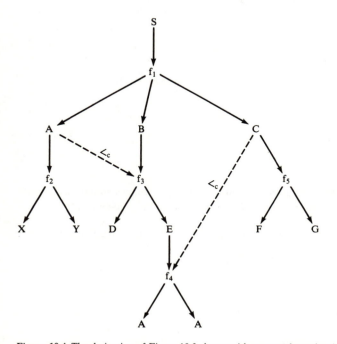

Figure 10.4 The derivation of Figure 10.3 shown with context dependencies.

The canonical derivation is the second one ($f_1 f_3 f_2 f_4 f_5$), and there are no other equivalent derivations.

The graphical representation of such derivation sequences is more involved than with the case of type 0 grammars, for there are three types of dependence which must be shown.

First, start with the derivation tree (with rewrite rules) in the underlying context-free grammar. For the example above, this is shown in Figure 10.3. Next, consider the dependencies of the instances of the rewrite rules on the context symbols. Figure 10.4 shows these dependencies for the example using directed arcs labeled with the symbol $<_c$ (for "context dependency"). Finally, a dependency, $<_f$, is shown between instances of rewrite rules to show that one rule is "free" to be applied after the other rule has used its rewritten symbol for context. The rule for the $<_c$ relation is as follows (where each n_i is

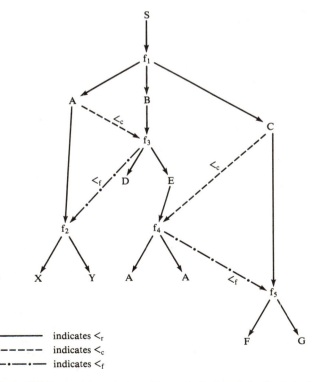

————— indicates $<_r$

- - - - - indicates $<_c$

—·—·— indicates $<_f$

Figure 10.5 A complete context-sensitive syntactical graph for the derivation in Figure 10.3.

a node of the graph):

If n_1 is labeled by $A \in V_N$ and n_2 is labeled by $f_2 \in F$ with $(n_2, n_1) \in <_c$ (that is, A is a context symbol for f_2), n_3 is labeled by $f_1 \in F$ where $f_1 : A \to P$, and f_1 is applied to n_1, then set $(n_3, n_2) \in <_f$.

Figure 10.5 on page 181 shows the final resulting *context-sensitive syntactical graph* (CSSG) for the example. "Rewrite dependency" is denoted by $<_r$, which represents the directed arcs of the original derivation tree.

Hart [1980] goes on to show that a CSSG forms a DOG (doubly ordered graph, see Section 10.4) when all three context-sensitive dependencies are formed. Then, in a nice analogy to Theorem 10.7, the total order $B \cup L^r$ on this DOG yields the rewrite rules exactly in canonical order.

EXERCISES

10.8 Give an inductive definition of the three context-sensitive relations: $<_r$, $<_c$, and $<_f$. Define the resulting B and L relations to give a DOG. Show that the $B \cup L^r$ total order yields the canonical derivation.

10.9 Theorem 4.1 shows how to convert a length increasing grammar into a context-sensitive grammar. Assuming that the original grammar is in Kuroda normal form (Definition 4.2), show the appropriate segments of the CSSG which follow from the conversion of a rule of the form $AB \to CD$.

APPENDIX

ELEMENTS OF SET THEORY

In this book we are using only the basic facts of naive set theory, which are summarized here for convenience. The notion of set is used in the intuitive sense as any collection (or heap) of arbitrary objects called elements. Therefore a set may also be an element of another set and we do not distinguish between sets and classes. (In axiomatic set theory classes represent very large sets which cannot be elements of sets or classes, to avoid paradoxes.)

A set is given if its elements are given. Thus, a finite set can be given by listing its elements in some (arbitrary) order. The notation

$$H = \{x_1, x_2, \ldots, x_n\}$$

is used to specify the finite set H having the elements $x_1, x_2, \ldots, x_n$. The set of natural numbers is usually denoted by N and is considered to be a well-defined infinite set. For any object x and set H the notation

$$x \in H$$

means that x *is an element of* H. The expressions H *contains* x, x *belongs to* H, x *is a member of* H, and x *is in* H are used synonymously with $x \in H$. The negation of $x \in H$ is denoted by $x \notin H$.

If a set H contains every element of K, then we say that K *is a subset of* H or H *includes* K; in symbols

$$K \subseteq H$$

Two sets are identical if they contain the same elements. Hence,

$$K = H$$

iff both $K \subseteq H$ and $H \subseteq K$. *Proper inclusion* is denoted by $K \subset H$, which means that $K \subseteq H$, but $K \neq H$.

A subset of a given set can be specified by some characteristic property. For instance, we can select those natural numbers which are powers of 2. For this purpose we use the notation

$$H = \{x \in N \mid x = 2^n \text{ with } n \in N\}$$

where N denotes the set of natural numbers. In general

$$K = \{x \in H \mid x \text{ has the property } P\}$$

denotes the set of those elements of H which have the property P. The vertical bar in this notation reads "such that." Note that in order for K to be well-defined, the set H must be known beforehand and the property P must be well-defined. For instance, the set

$$\{x \in N \mid x = 2^p \text{ with } p \text{ being a prime number}\}$$

which can be abbreviated as

$$\{2^p \mid p \text{ is a prime number}\}$$

is a well-defined set. So is the set

$$\{n \in N \mid \text{the regular polygon with } n \text{ sides is constructible}\}$$

although it is much more difficult to decide whether or not n has the property in question. On the other hand, it is unknown whether the set

$$\{n \in N \mid n > 2 \text{ and } x^n + y^n = z^n \text{ for some positive integers } x, y, \text{ and } z\}$$

contains any element at all. One can also find properties which are known to be undecidable.

For a well-defined property it can happen, of course, that none of the elements of the given set has that property. In this case the resulting set is empty. The *empty set* is denoted by $\varnothing$ and by definition $x \notin \varnothing$ for any object x.

A set is *finite* if it contains a finite number of elements, otherwise it is *infinite*. An infinite set is *countable* (or denumerable) iff its elements can be arranged in some sequence

$$x_1, x_2, \ldots$$

where each element occurs at least once. This means that a unique element of the set is assigned to each index value. (In other words, there exists a mapping (or function) of the natural numbers onto every countable set.)

Given two sets H and K, we define the following operations:

$$\text{Union: } H \cup K = \{x \mid x \in H \text{ or } x \in K\}$$

$$\text{Intersection: } H \cap K = \{x \mid x \in H \text{ and } x \in K\}$$

$$\text{Difference: } H - K = \{x \mid x \in H \text{ and } x \notin K\}$$

Complementation is a special case of the difference which makes sense only in the case when all the sets in question are subsets of some well-defined totality called the complete set. The *complete set* is usually denoted by I and for every $H \subseteq I$

$$\bar{H} = I - H$$

denotes the complement of H with respect to I.

It can be shown easily from the definitions that these set theoretic operations satisfy the following identities:

1) $H \cup H = H$ $H \cap H = H$

2) $H \cup K = K \cup H$ $H \cap K = K \cap H$

3) $H \cup (K \cup L) = (H \cup K) \cup L$ $H \cap (K \cap L) = (H \cap K) \cap L$

4) $H \cup (K \cap L) = (H \cup K)$ $H \cap (K \cup L) = (H \cap K) \cup (H \cap L)$
      $\cap (H \cup L)$

5) $H \cup (H \cap K) = H$ $H \cap (H \cup K) = H$

6) $H \cup \varnothing = H$ $H \cap \varnothing = \varnothing$

7) $H \cup I = I$ $H \cap I = H$

8) $H \cup \bar{H} = I$ $H \cap \bar{H} = \varnothing$

9) $\overline{(H \cup K)} = \bar{H} \cap \bar{K}$ $\overline{(H \cap K)} = \bar{H} \cup \bar{K}$

For instance, the first of the so-called De Morgan identities can be shown this way:

$$\overline{(H \cup K)} = \{x \,|\, x \notin (H \cup K)\} = \{x \,|\, x \notin H \text{ and } x \notin K\} = \bar{H} \cap \bar{K}$$

The set of all subsets of H is called the *power set of H* and is sometimes denoted by 2^H. (A finite set with n elements has obviously 2^n different subsets. It has $\binom{n}{k}$ different k-element subsets and $2^n = \binom{n}{0} + \binom{n}{1} + \cdots + \binom{n}{n}$.)

A family of sets forms a *boolean algebra* iff it is closed under the operations of $\cup$, $\cap$, and complementation. (By closure we mean that the result belongs to the same family.) For instance, the family

$$B = \{\varnothing, \{a\}, \{b\}, \{a, b\}\}$$

is a boolean algebra where $\{a, b\}$ corresponds to I. If we leave out one of the four sets, the remaining three is no longer a boolean algebra, but $\varnothing$ and I obviously form a boolean algebra. The power set of an arbitrary set is always a boolean algebra.

The operations of $\cup$ and $\cap$ are extended to any finite number of sets $H_1, \ldots, H_n$, namely

$$\bigcup_{i=1}^{n} H_i = \{x \,|\, x \in H_i \text{ for some } i \text{ with } 1 \leqslant i \leqslant n\}$$

and

$$\bigcap_{i=1}^{n} H_i = \{x \,|\, x \in H_i \text{ for every } i \text{ with } 1 \leqslant i \leqslant n\}$$

The set of *ordered pairs* of the form $[x, y]$ with $x \in H$ and $y \in K$ is called the *cartesian product* of H and K and is denoted by $H \times K$. In short,

$$H \times K = \{[x, y] \mid x \in H \text{ and } y \in K\}$$

This can also be extended to any finite sequence of operands

$$H_1 \times \cdots \times H_n = \{[x_1, \ldots, x_n] \mid x_i \in H_i \text{ for every } i \text{ with } 1 \leqslant i \leqslant n\}$$

where $[x_1, \ldots, x_n]$ is called an *ordered n-tuple*.

For an infinite sequence of sets $H_1, H_2, \ldots$ we define

$$\bigcup_{i=1}^{\infty} H_i = \{x \mid x \in H_i \text{ for some } i \in N\}$$

and

$$\bigcap_{i=1}^{\infty} H_i = \{x \mid x \in H_i \text{ for every } i \in N\}$$

The union of countably many countable sets is again countable which can be shown with the aid of the two-dimensional arrangement

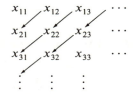

by rearranging its elements into a linear sequence following the diagonal arrows:

$$x_{11} \quad x_{12} \quad x_{21} \quad x_{13} \quad x_{22} \quad x_{31} \quad \cdots$$

The same method can be used to show that the cartesian product of two countable sets is again countable.

The power set of a countable set is, however, *noncountable*, which can be shown indirectly as follows. Suppose that the subsets of a countable set H can be arranged into a sequence

$$S_1, S_2, \ldots$$

where $S_i \subseteq H$ for every $i \in N$. Consider an arrangement

$$x_1, x_2, \ldots$$

of the elements of H. Divide the elements of H into two subsets, G and $H - G$, such that

$$x_i \in G \quad \text{iff } x_i \in S_i$$

Since both G and $H - G$ are subsets of H they must appear somewhere in the sequence $S_1, S_2, \ldots$. But if there is some $j \in N$ with

$$H - G = S_j$$

then $x_j \in S_j$ implies $x_j \notin H - G = S_j$ and conversely $x_j \notin S_j$ implies $x_j \in H - G = S_j$, which is a contradiction. Therefore, the assumption (the countability of the power set of H) must be false.

Mappings can be used in general to compare sets. *A mapping from a set H to a set K* is defined as a set of ordered pairs

$$M \subseteq H \times K$$

such that $[x, y] \in M$ and $[x, z] \in M$ implies $y = z$.

The *domain* of a mapping M is the set

$$dom(M) = \{x \in H | [x, y] \in M \text{ for some } y \in K\}$$

and the *range* or *codomain* of M is

$$cod(M) = \{y \in K | [x, y] \in M \text{ for some } x \in H\}$$

A mapping from H to K is *total* iff $dom(M) = H$ and it is *onto* iff $cod(M) = K$. A *total* mapping from H *onto* K is called a *mapping of H onto K*.

An arbitrary subset of $H \times H$ is called a *binary relation* on H. A binary relation R on H is *symmetric* iff $[x, y] \in R$ implies $[y, x] \in R$. It is *reflexive* iff $[x, x]$ is in R for every $x \in H$, and it is *transitive* iff $[x, y] \in R$ and $[y, z] \in R$ implies $[x, z] \in R$. A reflexive, symmetric, and transitive relation is called an *equivalence* relation. An equivalence relation on H defines a *partition* of H into disjoint subsets called *equivalence classes*. The *transitive closure* of R, denoted by R^+ is defined inductively as follows:

1) $[x, y] \in R^+$ for all $[x, y] \in R$
2) If $[x, y] \in R^+$ and $[y, z] \in R$, then $[x, z] \in R^+$

The *reflexive and transitive closure* of R is denoted by R^* and is defined as $R^* = R^+ \cup \{[x, x] | x \in H\}$.

We often use the conventional notation $m(x)$ for a mapping (or function) from H to K where $x \in H$ and $m(x) \in K$. Also, a binary relation on a set H is often written in the *infix notation* xRy rather than $[x, y] \in R$.

Every mapping from H to K has a *natural extension* to the subsets of H. Namely, for any subset $L \subseteq H$ we can define its image under the mapping m as the following subset of K

$$m(L) = \{y \in K | y = m(x) \text{ for some } x \in L\}$$

This way we get a mapping from 2^H to 2^K.

BIBLIOGRAPHIC NOTES

These notes serve two purposes. On one hand, they specify the sources of information used in this book; on the other hand, they give some ideas for further readings. They are by no means intended to be exhaustive.

There are three major books currently available on formal languages which cover more material than this book. The best known is perhaps Hopcroft and Ullman [1979] whose earlier version appeared in 1969. Salomaa [1973] is a fairly comprehensive survey of the subject at that time. Harrison [1978] gives exceptionally detailed treatment of deterministic languages. Each of these books includes extensive bibliographical and historical notes.

For a brief history of the theory of formal languages, see Greibach [1981]. An early text on context-free languages is Ginsburg [1966]. For the specific area of Lindenmayer systems, not discussed in our book, see Rozenberg and Salomaa [1978]. The rest of these notes will follow the order of the topics as presented chapter by chapter in our book.

Chapter 1. The idea of phrase-structure grammars and their basic hierarchy is certainly due to Chomsky [1956] and [1959]. The grammar in Example 1.2 is due to M. Soittola while our grammar in Exercise 1.5 is much simpler than his grammar for the same language. (See Examples 2.1 and 2.3 in Salomaa [1973].)

Chapter 2. Closure properties have been studied extensively by many people. Our proof of Theorem 2.2 follows Salomaa [1973] where references to original works can also be found. The algebraically oriented theory of abstract families of languages is summarized in Ginsburg [1975]. The idea of homomorphism comes also from algebra and, in particular, from the theory of groups and semigroups which are clearly related to the theory of automata and formal languages.

Chapter 3. Chomsky normal form appears in Chomsky [1959]. Derivation trees are already present in the earliest papers. The pumping lemma is from Bar-Hillel et al. [1961]. A stronger version of the pumping lemma can be found in Ogden [1968]. Greibach normal form appears in Greibach [1965]. Regular expressions were studied by Kleene [1956]. Our Theorem 3.12 is equivalent to his fundamental theorem appearing in that paper, but our proof follows the method of McNaughton and Yamada [1960]. For the axiom system of regular expressions see Salomaa [1966] and T. Urponen [1971].

Chapter 4. Kuroda normal form appears in Kuroda [1964]. One-sided context was shown to have the same generative power as two-sided context by Penttonen [1974a]. The proof of Theorem 4.4 is from Révész [1974] while Example 4.1 is from Révész [1971]. The fact that programming languages have context-sensitive features is pointed out by many authors.

Chapter 5. Theorem 5.1 is from Révész [1976]. Derivation graphs appear in Griffiths [1968] and Loeckx [1970]. Their string representation discussed in Chapter 10 is from Hart [1974].

Chapter 6. There are several books dealing with finite automata: Ginzburg [1968], Salomaa [1969], and Starke [1969] are some of the classics.

The originators of the subject appear to be McCulloch and Pitts [1943] who also had an impact on the logical design of the first von Neumann computer, the EDVAC. The foundations of the theory were laid down by Huffman [1954], Mealy [1955], Moore [1956], and Kleene [1956]. Nondeterministic finite automata were introduced by Rabin and Scott [1959] who also proved their equivalence to the deterministic model.

The equivalence of context-free grammars and pushdown automata was shown by Chomsky [1962] and Evey [1963]. Deterministic pushdown automata were studied by Fisher [1963], Schutzenberger [1963], Ginsburg and Greibach [1966], and many others. Their equivalence to $LR(k)$ grammars appears in Knuth [1965].

Turing machines were introduced by Turing [1936]. Chomsky [1953] proved their equivalence to phrase-structure grammars. Myhill [1960] defined deterministic linear bounded automata. The equivalence of nondeterministic linear bounded automata and context-sensitive grammars was shown by Kuroda [1964]. Our presentation of Sections 6.3 and 6.4 follows Révész [1976].

For the so-called *LBA* problem see Hartmanis and Hunt [1973]. Automata with output, such as general sequential machines (finite transducers) and others are not discussed in our book but they are straightforward extensions of the basic models.

Chapter 7. The theory of algorithms, i.e. the theory of effective computability and decidability was originally concerned with the foundations of mathematics. The appearance of electronic computers has obviously influenced the new development of this area. Davis [1958], Rogers [1967], and Minsky [1967] are important books on this subject. A more recent book is Machtey and Young [1978] which shows the shift of the emphasis towards practical computations. For a historical review see Kleene [1981]. The constructions of Theorems 7.9 and 7.10 are from Hartmanis [1967].

Chapter 8. Complexity theory can be traced back to Rabin [1960] and Yamada [1962]. The first decisive paper on time complexity is Hartmanis and Stearns [1965]. Space complexity was systematically studied first in Stearns, Hartmanis, and Lewis [1965]. Since then, complexity theory has made spectacular progress and still is one of the most active areas of theoretical computer science. A historical account of the early development is given by Hartmanis [1981]. The significance of nondeterministic models in case of polynomial boundary functions was first emphasized by Cook [1971] and Karp [1971].

Two excellent textbooks on the subject are Aho, Hopcroft, and Ullman [1974] and Baase [1978]. Somewhat more specialized is Borodin and Munro [1975]. Lewis and Papadimitriou [1981] cover a much broader area including formal languages and mathematical logic.

Theorem 8.4 is due to Savitch [1970]. Theorem 8.5 is based on a parsing algorithm found independently by Cocke, Kasami, and Younger. Our proof follows Graham and Harrison [1976]. The assymptotic improvement is given by Valiant [1975]. The hardest context-free language (except for its grammar) is due to Greibach [1973].

Chapter 9. The formal theory of the semantics has been lagging far behind the syntax of programming languages. Due to the inherent difficulty of the problems involved, no major breakthrough is to be expected soon. The practical need for efficient compiler design and program verification stimulates a growing interest in this area as can be seen, e.g., from the *Proceedings* edited by Jones [1980]. The three main directions of research have slightly different aims. Operational semantics is mainly concerned with compiler construction. The axiomatic approach makes use of mathematical logic (predicate calculus) to describe and prove properties of programs. Denotational semantics is concerned with abstract mathematical definitions, mainly for language designers.

Attributed grammars were introduced by Knuth [1968]. The axiomatic approach was introduced by Hoare [1969]. Donahue [1976], Stoy [1977], and Gordon [1979] may be used as introductory texts with emphasis on the denotational approach.

Ambiguity and inherent ambiguity is discussed in more detail by Hopcroft and Ullman [1979]. Our treatment follows Takaoka [1974]. More on formal power series can be found in Salomaa and Soittola [1978].

Earley's algorithm appears in Earley [1970]. $LL(k)$ grammars were introduced by Lewis and Stearns [1968]. The corresponding language hierarchy was established by Kurki-Suonio [1969]. Our treatment of *LF* grammars is a simplification of the original presentation by Wood [1969 and 1970]. *LR(k)* grammars were introduced by Knuth [1965]. More on syntax analysis (or parsing) may be found in books dealing with compiler construction like Lewis, Rosenkrantz, and Stearns [1976], Aho and Ullman [1979], Barret and Couch [1979], and Pyster [1981].

Chapter 10. Abstract algebraic properties of derivations in phrase structure grammars were studied by Hotz [1966]. The string representations called derivation words were introduced by Hart [1974]. The self-embedding property expressed by Theorem 10.1 is from Révész [1977]. Sections 10.2 and 10.3 follow the same paper while the rest of Chapter 10 is based on Hart [1976 and 1980].

REFERENCES

Aho, A. V., J. E. Hopcroft, and J. D. Ullman: *The Design and Analysis of Computer Algorithms*, Addison-Wesley, Reading, Mass., 1974.

_____, and J. D. Ullman: *Principles of Compiler Design*, Addison-Wesley, Reading, Mass., 1977.

Baase, Sara: *Computer Algorithms: Introduction to Design and Analysis*, Addison-Wesley, Reading, Mass., 1978.

Bar-Hillel, Y., M. Perles, and E. Shamir: "On formal properties of simple phrase structure grammars," *Zeitschrift fur Phonetik, Sprachwisenschaft, und Kommunikationsforschung, 14* (1961), pp. 143–172.

Barret, W. A., and J. D. Couch: *Compiler Construction: Theory and Practice*, Science Research Associates, Inc., 1979.

Borodin, A., and I. Munro: *The Computational Complexity of Algebraic and Numeric Problems*, American Elsevier, New York, 1975.

Chomsky, N.: "Three models for the description of language," *IRE Transactions on Information Theory, 2*:3 (1956), pp. 113–124.

_____: "On certain formal properties of grammars," *Information and Control, 2* (1959), pp. 137–167.

_____: "Context-free grammars and pushdown storage," *MIT Research Laboratory of Electronics, Quarterly Progress Report*, no. 65 (1962), pp. 187–194.

Cook, S. A.: "The complexity of theorem-proving procedures," *Proceedings of the Third Annual ACM Symposium on Theory of Computing* (1971), pp. 151–158.

Davis, M.: *Computability and Unsolvability*, McGraw-Hill, N.Y., 1958.

Donahue, J. E.: "Complementary definitions of programming language semantics," *Lecture Notes in Computer Science*, vol. 42, Springer-Verlag, New York, 1976.

Earley, J.: "An efficient context-free parsing algorithm," *Communications of the ACM, 13* (1956), pp. 94–102.

Evey, R. J.: "Application of pushdown-store machines," *Proceedings 1963 Fall Joint Computer Conference*, AFIPS Press, 1963, pp. 215–227.

Fisher, P.C.: "On computability by certain classes of restricted Turing machines," *Proceedings Fourth Annual IEEE Symposium on Switching Circuit Theory and Logical Design*, (1963), 23–32.

Ginsburg, S.: *The Mathematical Theory of Context-Free Languages*, McGraw-Hill, N.Y., 1966.

Ginsburg, S., and S. Griebach: "Deterministic context-free languages," *Information & Control, 9* (1966), pp. 620–648.

Ginsburg, S., and E. Spanier: "Control sets on grammars," *Mathematical Systems Theory, 2* (1968), pp. 159–177.

Ginzburg, Abraham: *Algebraic Theory of Automata*, Academic Press, N.Y., 1968.

Gordon, M. J. C.: *The Denotational Description of Programming Languages*, Springer-Verlag, 1979.

Graham, S. L., and M. A. Harrison: "Parsing of general context-free languages," *Advances in Computers, 14* (1976), Academic Press, N.Y., pp. 77–185.

Greibach, S.: "A new normal-form theorem for context-free phrase structure grammars," *Journal of the ACM, 12* (1965), pp. 42–52.

_____: "The hardest context-free language," *SIAM Journal on Computing, 2* (1973), pp. 304–310.

_____: "Formal languages: Origins and directions," *Annals of the History of Computing, 3,* 1 (1981), pp. 14–41.

Griffiths, T. V.: "Some remarks on derivations in general rewriting systems," *Information & Control, 12* (1968), pp. 27–54.

Harrison, M. A.: *Introduction to formal language theory*, Addison-Wesley, Reading, Mass., 1978.

Hart, J. M.: "Acceptors for the derivation languages of phrase-structure grammars," *Information & Control, 25* (1974), pp. 75–92.

_____: "Right and left parses in phrase-structure grammars," *Information & Control, 32* (1976), pp. 242–262.

_____: "Derivation structures for strictly context-sensitive grammars," *Information & Control, 45* (1980), pp. 68–89.

Hartmanis, J.: "Context-free languages and Turing machine computations," *Proceedings Symposium on Applied Mathematics, American Mathematical Society, 19* (1967), pp. 42–51.

_____: "Observations about the development of theoretical computer science," *Annals of the History of Computing, 3:* 1 (1981), pp. 42–51.

_____, and R. E. Stearns: "On the computational complexity of algorithms," *Transactions American Mathematical Society, 117* (1965), pp. 283–306.

_____, and H. B. Hunt, III: "The lba problem and its importance in the theory of computing," *Complexity of Computations* (R. Karp, ed.), *SIAM-AMS Proceedings, 8* (1973), pp. 27–42.

Hoare, C. A. R.: "An axiomatic basis for computer programming," *Communications of the ACM, 12* (1969), pp. 576–581.

Hopcroft, J. E., and J. D. Ullman: *Introduction to Automata Theory, Languages, and Computation*, Addison-Wesley, Reading, Mass., 1979.

Hotz, G.: "Eindeutigkeit und Mehrdeutigkeit formaler Sprachen," *Elektron. Informationsverarbeit. Kybernetik, 2* (1966), pp. 235–246.

Huffman, D. A.: "The synthesis of sequential switching networks," *Journal of the Franklin Institute, 257* (1954), pp. 161–190, 275–303.

Jones, N. D. (ed.) "Semantics-directed compiler generation," *Lecture Notes in Computer Science, 94* (1980), Springer-Verlag.

Karp, R. M.: "Reducibility among combinatorial problems," in *Complexity of Computations*, Plenum Press, N. Y., 1972.

Kleene, S. C.: "Representation of events in nerve nets," in *Automata Studies*, Princeton University Press, Princeton, N.J., 1956.

_____: "Origins of recursive function theory," *Annals of the History of Computing, 3:*1 (1981), pp. 52–67.

Knuth, D. E.: "On the translation of languages from left to right," *Information & Control, 8* (1965), pp. 607–633.

_____: "Semantics of context-free languages," *Mathematical Systems Theory, 2:*2 (1968), pp. 127–145.

Kuroda, S. Y.: "Classes of languages and linear-bounded automata," *Information & Control*, 7 (1964), 207–223.

Kurki-Suonio, R.: "Notes on top-down languages," *BIT*, 9 (1969), pp. 225–238.

Lewis, H. R., and C. H. Papadimitriou: *Elements of the Theory of Computation*, Prentice-Hall, Inc., Englewood Cliffs, N.J., 1981.

Lewis, P. M., II, D. J. Rosenkrantz, and R. E. Stearns: *Compiler Design Theory*, Addison-Wesley, Reading, Mass., 1976.

Loeckx, J.: "The parsing of general phrase structure grammars," *Information & Control*, 16 (1970), pp. 443–464.

Machtey, M., and P. Young: *An Introduction to the General Theory of Algorithms*, North-Holland, N.Y., 1978.

McCullough, W. S., and E. Pitts: "A logical calculus of the ideas immanent in nervous activity," *Bulletin of Mathematical Biophysics*, 5 (1943), pp. 115–133.

McNaughton, R., and H. Yamada: "Regular expressions and state graphs for automata," *IRE Transactions on Electronic Computers*, 9:1 (1960), pp. 39–47.

Mealy, G. H.: "Method for synthesizing sequential circuits," *Bell System Technical Journal*, 34 (1955), pp. 1054–1079.

Minsky, M.: *Computation: Finite and Infinite Machines*, Prentice Hall, Englewood Cliffs, N.J., 1967.

Moore, E. F.: "Gedanken-Experiments on sequential machines," in *Automata Studies*, Princeton University Press, Princeton, N.J., 1956, pp. 129–153.

Myhill, J.: *Linear Bounded Automata*, WADD TR-60-165, Ohio, 1960, pp. 60–165.

Ogden, W.: "A helpful result for proving inherent ambiguity," *Mathematical Systems Theory*, 2 (1968), pp. 191–194.

Penttonen, M.: "One-sided and two-sided context in formal grammars," *Information & Control*, 25 (1974a), pp. 371–392.

_____: "On derivation languages corresponding to context-free grammars," *Acta Informatica*, 3 (1974b), pp. 285–291.

Pyster, A. B.: *Compiler Design and Construction*, Van Nostrand Reinhold, 1981.

Rabin, M. O.: "Degree of difficulty of computing a function and a partial ordering of recursive sets," *Technical Report No. 3*, Hebrew University, Jerusalem, 1960.

_____, and D. Scott: "Finite automata and their decision problems," *IBM Journal of Research and Development*, 3 (1959), pp. 114–125.

Révész, G.: "Unilateral context sensitive grammars and left-to-right parsing," *Journal of Computer and System Sciences*, 5 (1971), pp. 337–352.

_____: "Comment on the paper 'Error detection in formal languages'," *Journal of Computer and System Sciences*, 8 (1974), pp. 238–242.

_____: "A note on the relation of Turing machines to phrase structure grammars," *Computational Linguistics and Computer Languages*, 11 (1976), pp. 11–16.

_____: "Algebraic properties of derivation words," *Journal of Computer and System Sciences*, 15 (1977), pp. 232–240.

Rogers, H., Jr.: *The Theory of Recursive Functions and Effective Computability*, McGraw-Hill, N.Y., 1967.

Rozenberg, G., and A. Salomaa: "L Systems," *Lecture Notes in Computer Science*, No. 15, Springer-Verlag, Berlin, 1978.

Salomaa, A.: *Theory of Automata*, Pergamon Press, Oxford, 1969.

_____: *Formal Languages*, Academic Press, N.Y., 1973.

_____: "Two complete axiom systems for the algebra of regular events," *Journal of the ACM*, 13 (1966), pp. 158–163.

_____, and M. Soittola: *Automata-Theoretic Aspects of Formal Power Series*, Springer-Verlag, New York, 1978.

Savitch, W. J.: "Relationship between nondeterministic and deterministic tape complexities," *Journal of Computer and System Sciences, 4* (1970), pp. 177–192.

Schutzenberger, M. P.: "On context-free languages and pushdown automata," *Information & Control, 6* (1963), pp. 246–264.

Stearns, R. E., J. Hartmanis, and P. M. Lewis, II: "Hierarchies of memory limited computations," *IEEE Conference Record on Switching Circuit Theory and Logical Design,* (1965), pp. 179–190.

Stoy, J.: *Denotational Semantics: The Scott-Strachey Approach to Programming Language Theory,* MIT Press, Cambridge, Mass., 1977.

Takaoka, T.: "A note on the ambiguity of context-free grammars," *Information Processing Letters, 3*:2 (1974), pp. 34–36.

Turing, A. M.: "On computable numbers, with an application to the Entscheidungsproblem," *Proceedings London Mathematical Society, 42* (1936), pp. 230–265, *43* (1936), pp. 544–546.

Urponen, T.: "On axiom systems for regular expressions and on equations involving languages," *Ann. Univ. Turku,* Ser. AI 145 (1971).

Valiant, L. G.: "General context-free recognition in less than cubic time," *Journal of Computer and System Sciences, 10* (1975), pp. 308–315.

Wood, D.: "The theory of left-factored languages," *The Computer Journal, 12*:4 (1969), pp. 349–356, *13*:1 (1970), pp. 55–62.

Yamada, H.: "Real-time computation and recursive functions not real-time computable," *IRE Transactions on Electronic Computers,* EC11, 1962, pp. 753–760.

INDEX

A CATALOG OF SELECTED
DOVER BOOKS
IN SCIENCE AND MATHEMATICS

A CATALOG OF SELECTED
DOVER BOOKS
IN SCIENCE AND MATHEMATICS

QUALITATIVE THEORY OF DIFFERENTIAL EQUATIONS, V.V. Nemytskii and V.V. Stepanov. Classic graduate-level text by two prominent Soviet mathematicians covers classical differential equations as well as topological dynamics and erqodic theory. Bibliographies. 523pp. 5⅜ × 8½. 65954-2 Pa. $10.95

MATRICES AND LINEAR ALGEBRA, Hans Schneider and George Phillip Barker. Basic textbook covers theory of matrices and its applications to systems of linear equations and related topics such as determinants, eigenvalues and differential equations. Numerous exercises. 432pp. 5⅜ × 8½. 66014-1 Pa. $8.95

QUANTUM THEORY, David Bohm. This advanced undergraduate-level text presents the quantum theory in terms of qualitative and imaginative concepts, followed by specific applications worked out in mathematical detail. Preface. Index. 655pp. 5⅜ × 8½. 65969-0 Pa. $10.95

ATOMIC PHYSICS (8th edition), Max Born. Nobel laureate's lucid treatment of kinetic theory of gases, elementary particles, nuclear atom, wave-corpuscles, atomic structure and spectral lines, much more. Over 40 appendices, bibliography. 495pp. 5⅜ × 8½. 65984-4 Pa. $11.95

ELECTRONIC STRUCTURE AND THE PROPERTIES OF SOLIDS: The Physics of the Chemical Bond, Walter A. Harrison. Innovative text offers basic understanding of the electronic structure of covalent and ionic solids, simple metals, transition metals and their compounds. Problems. 1980 edition. 582pp. 6⅛ × 9¼. 66021-4 Pa. $14.95

BOUNDARY VALUE PROBLEMS OF HEAT CONDUCTION, M. Necati Özisik. Systematic, comprehensive treatment of modern mathematical methods of solving problems in heat conduction and diffusion. Numerous examples and problems. Selected references. Appendices. 505pp. 5⅜ × 8½. 65990-9 Pa. $11.95

A SHORT HISTORY OF CHEMISTRY (3rd edition), J.R. Partington. Classic exposition explores origins of chemistry, alchemy, early medical chemistry, nature of atmosphere, theory of valency, laws and structure of atomic theory, much more. 428pp. 5⅜ × 8½. (Available in U.S. only) 65977-1 Pa. $10.95

A HISTORY OF ASTRONOMY, A. Pannekoek. Well-balanced, carefully reasoned study covers such topics as Ptolemaic theory, work of Copernicus, Kepler, Newton, Eddington's work on stars, much more. Illustrated. References. 521pp. 5⅜ × 8½. 65994-1 Pa. $11.95

PRINCIPLES OF METEOROLOGICAL ANALYSIS, Walter J. Saucier. Highly respected, abundantly illustrated classic reviews atmospheric variables, hydrostatics, static stability, various analyses (scalar, cross-section, isobaric, isentropic, more). For intermediate meteorology students. 454pp. 6⅛ × 9¼. 65979-8 Pa. $12.95

RELATIVITY, THERMODYNAMICS AND COSMOLOGY, Richard C. Tolman. Landmark study extends thermodynamics to special, general relativity; also applications of relativistic mechanics, thermodynamics to cosmological models. 501pp. 5⅜ × 8½. 65383-8 Pa. $11.95

APPLIED ANALYSIS, Cornelius Lanczos. Classic work on analysis and design of finite processes for approximating solution of analytical problems. Algebraic equations, matrices, harmonic analysis, quadrature methods, much more. 559pp. 5⅜ × 8½. 65656-X Pa. $11.95

SPECIAL RELATIVITY FOR PHYSICISTS, G. Stephenson and C.W. Kilmister. Concise elegant account for nonspecialists. Lorentz transformation, optical and dynamical applications, more. Bibliography. 108pp. 5⅜ × 8½. 65519-9 Pa. $3.95

INTRODUCTION TO ANALYSIS, Maxwell Rosenlicht. Unusually clear, accessible coverage of set theory, real number system, metric spaces, continuous functions, Riemann integration, multiple integrals, more. Wide range of problems. Undergraduate level. Bibliography. 254pp. 5⅜ × 8½. 65038-3 Pa. $7.00

INTRODUCTION TO QUANTUM MECHANICS With Applications to Chemistry, Linus Pauling & E. Bright Wilson, Jr. Classic undergraduate text by Nobel Prize winner applies quantum mechanics to chemical and physical problems. Numerous tables and figures enhance the text. Chapter bibliographies. Appendices. Index. 468pp. 5⅜ × 8½. 64871-0 Pa. $9.95

ASYMPTOTIC EXPANSIONS OF INTEGRALS, Norman Bleistein & Richard A. Handelsman. Best introduction to important field with applications in a variety of scientific disciplines. New preface. Problems. Diagrams. Tables. Bibliography. Index. 448pp. 5⅜ × 8½. 65082-0 Pa. $10.95

MATHEMATICS APPLIED TO CONTINUUM MECHANICS, Lee A. Segel. Analyzes models of fluid flow and solid deformation. For upper-level math, science and engineering students. 608pp. 5⅜ × 8½. 65369-2 Pa. $12.95

ELEMENTS OF REAL ANALYSIS, David A. Sprecher. Classic text covers fundamental concepts, real number system, point sets, functions of a real variable, Fourier series, much more. Over 500 exercises. 352pp. 5⅜ × 8½. 65385-4 Pa. $8.95

PHYSICAL PRINCIPLES OF THE QUANTUM THEORY, Werner Heisenberg. Nobel Laureate discusses quantum theory, uncertainty, wave mechanics, work of Dirac, Schroedinger, Compton, Wilson, Einstein, etc. 184pp. 5⅜ × 8½. 60113-7 Pa. $4.95

INTRODUCTORY REAL ANALYSIS, A.N. Kolmogorov, S.V. Fomin. Translated by Richard A. Silverman. Self-contained, evenly paced introduction to real and functional analysis. Some 350 problems. 403pp. 5⅜ × 8½. 61226-0 Pa. $7.95

PROBLEMS AND SOLUTIONS IN QUANTUM CHEMISTRY AND PHYSICS, Charles S. Johnson, Jr. and Lee G. Pedersen. Unusually varied problems, detailed solutions in coverage of quantum mechanics, wave mechanics, angular momentum, molecular spectroscopy, scattering theory, more. 280 problems plus 139 supplementary exercises. 430pp. 6½ × 9¼. 65236-X Pa. $10.95

ASYMPTOTIC METHODS IN ANALYSIS, N.G. de Bruijn. An inexpensive, comprehensive guide to asymptotic methods—the pioneering work that teaches by explaining worked examples in detail. Index. 224pp. 5⅜ × 8½.   64221-6 Pa. $5.95

OPTICAL RESONANCE AND TWO-LEVEL ATOMS, L. Allen and J.H. Eberly. Clear, comprehensive introduction to basic principles behind all quantum optical resonance phenomena. 53 illustrations. Preface. Index. 256pp. 5⅜ × 8½.
65533-4 Pa. $6.95

COMPLEX VARIABLES, Francis J. Flanigan. Unusual approach, delaying complex algebra till harmonic functions have been analyzed from real variable viewpoint. Includes problems with answers. 364pp. 5⅜ × 8½.   61388-7 Pa. $7.95

ATOMIC SPECTRA AND ATOMIC STRUCTURE, Gerhard Herzberg. One of best introductions; especially for specialist in other fields. Treatment is physical rather than mathematical. 80 illustrations. 257pp. 5⅜ × 8½.   60115-3 Pa. $4.95

APPLIED COMPLEX VARIABLES, John W. Dettman. Step-by-step coverage of fundamentals of analytic function theory—plus lucid exposition of 5 important applications: Potential Theory; Ordinary Differential Equations; Fourier Transforms; Laplace Transforms; Asymptotic Expansions. 66 figures. Exercises at chapter ends. 512pp. 5⅜ × 8½.   64670-X Pa. $10.95

ULTRASONIC ABSORPTION: An Introduction to the Theory of Sound Absorption and Dispersion in Gases, Liquids and Solids, A.B. Bhatia. Standard reference in the field provides a clear, systematically organized introductory review of fundamental concepts for advanced graduate students, research workers. Numerous diagrams. Bibliography. 440pp. 5⅜ × 8½.   64917-2 Pa. $8.95

UNBOUNDED LINEAR OPERATORS: Theory and Applications, Seymour Goldberg. Classic presents systematic treatment of the theory of unbounded linear operators in normed linear spaces with applications to differential equations. Bibliography. 199pp. 5⅜ × 8½.   64830-3 Pa. $7.00

LIGHT SCATTERING BY SMALL PARTICLES, H.C. van de Hulst. Comprehensive treatment including full range of useful approximation methods for researchers in chemistry, meteorology and astronomy. 44 illustrations. 470pp. 5⅜ × 8½.   64228-3 Pa. $9.95

CONFORMAL MAPPING ON RIEMANN SURFACES, Harvey Cohn. Lucid, insightful book presents ideal coverage of subject. 334 exercises make book perfect for self-study. 55 figures. 352pp. 5⅜ × 8¼.   64025-6 Pa. $8.95

OPTICKS, Sir Isaac Newton. Newton's own experiments with spectroscopy, colors, lenses, reflection, refraction, etc., in language the layman can follow. Foreword by Albert Einstein. 532pp. 5⅜ × 8½.   60205-2 Pa. $8.95

GENERALIZED INTEGRAL TRANSFORMATIONS, A.H. Zemanian. Graduate-level study of recent generalizations of the Laplace, Mellin, Hankel, K. Weierstrass, convolution and other simple transformations. Bibliography. 320pp. 5⅜ × 8½.   65375-7 Pa. $7.95

THE ELECTROMAGNETIC FIELD, Albert Shadowitz. Comprehensive undergraduate text covers basics of electric and magnetic fields, builds up to electromagnetic theory. Also related topics, including relativity. Over 900 problems. 768pp. 5⅜ × 8¼. 65660-8 Pa. $15.95

FOURIER SERIES, Georgi P. Tolstov. Translated by Richard A. Silverman. A valuable addition to the literature on the subject, moving clearly from subject to subject and theorem to theorem. 107 problems, answers. 336pp. 5⅜ × 8½. 63317-9 Pa. $7.95

THEORY OF ELECTROMAGNETIC WAVE PROPAGATION, Charles Herach Papas. Graduate-level study discusses the Maxwell field equations, radiation from wire antennas, the Doppler effect and more. xiii + 244pp. 5⅜ × 8½. 65678-0 Pa. $6.95

DISTRIBUTION THEORY AND TRANSFORM ANALYSIS: An Introduction to Generalized Functions, with Applications, A.H. Zemanian. Provides basics of distribution theory, describes generalized Fourier and Laplace transformations. Numerous problems. 384pp. 5⅜ × 8½. 65479-6 Pa. $8.95

THE PHYSICS OF WAVES, William C. Elmore and Mark A. Heald. Unique overview of classical wave theory. Acoustics, optics, electromagnetic radiation, more. Ideal as classroom text or for self-study. Problems. 477pp. 5⅜ × 8½. 64926-1 Pa. $10.95

CALCULUS OF VARIATIONS WITH APPLICATIONS, George M. Ewing. Applications-oriented introduction to variational theory develops insight and promotes understanding of specialized books, research papers. Suitable for advanced undergraduate/graduate students as primary, supplementary text. 352pp. 5⅜ × 8½. 64856-7 Pa. $8.50

A TREATISE ON ELECTRICITY AND MAGNETISM, James Clerk Maxwell. Important foundation work of modern physics. Brings to final form Maxwell's theory of electromagnetism and rigorously derives his general equations of field theory. 1,084pp. 5⅜ × 8½. 60636-8, 60637-6 Pa., Two-vol. set $19.00

AN INTRODUCTION TO THE CALCULUS OF VARIATIONS, Charles Fox. Graduate-level text covers variations of an integral, isoperimetrical problems, least action, special relativity, approximations, more. References. 279pp. 5⅜ × 8½. 65499-0 Pa. $6.95

HYDRODYNAMIC AND HYDROMAGNETIC STABILITY, S. Chandrasekhar. Lucid examination of the Rayleigh-Benard problem; clear coverage of the theory of instabilities causing convection. 704pp. 5⅜ × 8¼. 64071-X Pa. $12.95

CALCULUS OF VARIATIONS, Robert Weinstock. Basic introduction covering isoperimetric problems, theory of elasticity, quantum mechanics, electrostatics, etc. Exercises throughout. 326pp. 5⅜ × 8½. 63069-2 Pa. $7.95

DYNAMICS OF FLUIDS IN POROUS MEDIA, Jacob Bear. For advanced students of ground water hydrology, soil mechanics and physics, drainage and irrigation engineering and more. 335 illustrations. Exercises, with answers. 784pp. 6⅛ × 9¼. 65675-6 Pa. $19.95

NUMERICAL METHODS FOR SCIENTISTS AND ENGINEERS, Richard Hamming. Classic text stresses frequency approach in coverage of algorithms, polynomial approximation, Fourier approximation, exponential approximation, other topics. Revised and enlarged 2nd edition. 721pp. 5⅜ × 8½.
65241-6 Pa. $14.95

THEORETICAL SOLID STATE PHYSICS, Vol. I: Perfect Lattices in Equilibrium; Vol. II: Non-Equilibrium and Disorder, William Jones and Norman H. March. Monumental reference work covers fundamental theory of equilibrium properties of perfect crystalline solids, non-equilibrium properties, defects and disordered systems. Appendices. Problems. Preface. Diagrams. Index. Bibliography. Total of 1,301pp. 5⅜ × 8½. Two volumes.
Vol. I 65015-4 Pa. $12.95
Vol. II 65016-2 Pa. $12.95

OPTIMIZATION THEORY WITH APPLICATIONS, Donald A. Pierre. Broadspectrum approach to important topic. Classical theory of minima and maxima, calculus of variations, simplex technique and linear programming, more. Many problems, examples. 640pp. 5⅜ × 8½.
65205-X Pa. $12.95

THE MODERN THEORY OF SOLIDS, Frederick Seitz. First inexpensive edition of classic work on theory of ionic crystals, free-electron theory of metals and semiconductors, molecular binding, much more. 736pp. 5⅜ × 8½.
65482-6 Pa. $14.95

ESSAYS ON THE THEORY OF NUMBERS, Richard Dedekind. Two classic essays by great German mathematician: on the theory of irrational numbers; and on transfinite numbers and properties of natural numbers. 115pp. 5⅜ × 8½.
21010-3 Pa. $4.95

THE FUNCTIONS OF MATHEMATICAL PHYSICS, Harry Hochstadt. Comprehensive treatment of orthogonal polynomials, hypergeometric functions, Hill's equation, much more. Bibliography. Index. 322pp. 5⅜ × 8½.
65214-9 Pa. $8.95

NUMBER THEORY AND ITS HISTORY, Oystein Ore. Unusually clear, accessible introduction covers counting, properties of numbers, prime numbers, much more. Bibliography. 380pp. 5⅜ × 8½.
65620-9 Pa. $8.95

THE VARIATIONAL PRINCIPLES OF MECHANICS, Cornelius Lanczos. Graduate level coverage of calculus of variations, equations of motion, relativistic mechanics, more. First inexpensive paperbound edition of classic treatise. Index. Bibliography. 418pp. 5⅜ × 8½.
65067-7 Pa. $10.95

MATHEMATICAL TABLES AND FORMULAS, Robert D. Carmichael and Edwin R. Smith. Logarithms, sines, tangents, trig functions, powers, roots, reciprocals, exponential and hyperbolic functions, formulas and theorems. 269pp. 5⅜ × 8½.
60111-0 Pa. $5.95

THEORETICAL PHYSICS, Georg Joos, with Ira M. Freeman. Classic overview covers essential math, mechanics, electromagnetic theory, thermodynamics, quantum mechanics, nuclear physics, other topics. First paperback edition. xxiii + 885pp. 5⅜ × 8½.
65227-0 Pa. $17.95

CATALOG OF DOVER BOOKS

HANDBOOK OF MATHEMATICAL FUNCTIONS WITH FORMULAS, GRAPHS, AND MATHEMATICAL TABLES, edited by Milton Abramowitz and Irene A. Stegun. Vast compendium: 29 sets of tables, some to as high as 20 places. 1,046pp. 8 × 10½. 61272-4 Pa. $21.95

MATHEMATICAL METHODS IN PHYSICS AND ENGINEERING, John W. Dettman. Algebraically based approach to vectors, mapping, diffraction, other topics in applied math. Also generalized functions, analytic function theory, more. Exercises. 448pp. 5⅜ × 8¼. 65649-7 Pa. $8.95

A SURVEY OF NUMERICAL MATHEMATICS, David M. Young and Robert Todd Gregory. Broad self-contained coverage of computer-oriented numerical algorithms for solving various types of mathematical problems in linear algebra, ordinary and partial, differential equations, much more. Exercises. Total of 1,248pp. 5⅜ × 8½. Two volumes. Vol. I 65691-8 Pa. $13.95
Vol. II 65692-6 Pa. $13.95

TENSOR ANALYSIS FOR PHYSICISTS, J.A. Schouten. Concise exposition of the mathematical basis of tensor analysis, integrated with well-chosen physical examples of the theory. Exercises. Index. Bibliography. 289pp. 5⅜ × 8½.
65582-2 Pa. $7.95

INTRODUCTION TO NUMERICAL ANALYSIS (2nd Edition), F.B. Hildebrand. Classic, fundamental treatment covers computation, approximation, interpolation, numerical differentiation and integration, other topics. 150 new problems. 669pp. 5⅜ × 8½. 65363-3 Pa. $13.95

INVESTIGATIONS ON THE THEORY OF THE BROWNIAN MOVEMENT, Albert Einstein. Five papers (1905–8) investigating dynamics of Brownian motion and evolving elementary theory. Notes by R. Fürth. 122pp. 5⅜ × 8½.
60304-0 Pa. $3.95

NUMERICAL METHODS FOR SCIENTISTS AND ENGINEERS, Richard Hamming. Classic text stresses frequency approach in coverage of algorithms, polynomial approximation, Fourier approximation, exponential approximation, other topics. Revised and enlarged 2nd edition. 721pp. 5⅜ × 8½. 65241-6 Pa. $14.95

AN INTRODUCTION TO STATISTICAL THERMODYNAMICS, Terrell L. Hill. Excellent basic text offers wide-ranging coverage of quantum statistical mechanics, systems of interacting molecules, quantum statistics, more. 523pp. 5⅜ × 8½. 65242-4 Pa. $10.95

ELEMENTARY DIFFERENTIAL EQUATIONS, William Ted Martin and Eric Reissner. Exceptionally, clear comprehensive introduction at undergraduate level. Nature and origin of differential equations, differential equations of first, second and higher orders. Picard's Theorem, much more. Problems with solutions. 331pp. 5⅜ × 8½. 65024-3 Pa. $8.95

STATISTICAL PHYSICS, Gregory H. Wannier. Classic text combines thermodynamics, statistical mechanics and kinetic theory in one unified presentation of thermal physics. Problems with solutions. Bibliography. 532pp. 5⅜ × 8½.
65401-X Pa. $10.95

ORDINARY DIFFERENTIAL EQUATIONS, Morris Tenenbaum and Harry Pollard. Exhaustive survey of ordinary differential equations for undergraduates in mathematics, engineering, science. Thorough analysis of theorems. Diagrams. Bibliography. Index. 818pp. 5⅜ × 8½. 64940-7 Pa. $15.95

STATISTICAL MECHANICS: Principles and Applications, Terrell L. Hill. Standard text covers fundamentals of statistical mechanics, applications to fluctuation theory, imperfect gases, distribution functions, more. 448pp. 5⅜ × 8½. 65390-0 Pa. $9.95

ORDINARY DIFFERENTIAL EQUATIONS AND STABILITY THEORY: An Introduction, David A. Sánchez. Brief, modern treatment. Linear equation, stability theory for autonomous and nonautonomous systems, etc. 164pp. 5⅜ × 8¼. 63828-6 Pa. $4.95

THIRTY YEARS THAT SHOOK PHYSICS: The Story of Quantum Theory, George Gamow. Lucid, accessible introduction to influential theory of energy and matter. Careful explanations of Dirac's anti-particles, Bohr's model of the atom, much more. 12 plates. Numerous drawings. 240pp. 5⅜ × 8½. 24895-X Pa. $5.95

ORDINARY DIFFERENTIAL EQUATIONS, I.G. Petrovski. Covers basic concepts, some differential equations and such aspects of the general theory as Euler lines, Arzel's theorem, Peano's existence theorem, Osgood's uniqueness theorem, more. 45 figures. Problems. Bibliography. Index. xi + 232pp. 5⅜ × 8½. 64683-1 Pa. $6.00

GREAT EXPERIMENTS IN PHYSICS: Firsthand Accounts from Galileo to Einstein, edited by Morris H. Shamos. 25 crucial discoveries: Newton's laws of motion, Chadwick's study of the neutron, Hertz on electromagnetic waves, more. Original accounts clearly annotated. 370pp. 5⅜ × 8½. 25346-5 Pa. $8.95

INTRODUCTION TO PARTIAL DIFFERENTIAL EQUATIONS WITH AP-PLICATIONS, E.C. Zachmanoglou and Dale W. Thoe. Essentials of partial differential equations applied to common problems in engineering and the physical sciences. Problems and answers. 416pp. 5⅜ × 8½. 65251-3 Pa. $9.95

BURNHAM'S CELESTIAL HANDBOOK, Robert Burnham, Jr. Thorough guide to the stars beyond our solar system. Exhaustive treatment. Alphabetical by constellation: Andromeda to Cetus in Vol. 1; Chamaeleon to Orion in Vol. 2; and Pavo to Vulpecula in Vol. 3. Hundreds of illustrations. Index in Vol. 3. 2,000pp. 6⅛ × 9¼. 23567-X, 23568-8, 23673-0 Pa., Three-vol. set $38.85

ASYMPTOTIC EXPANSIONS FOR ORDINARY DIFFERENTIAL EQUA-TIONS, Wolfgang Wasow. Outstanding text covers asymptotic power series, Jordan's canonical form, turning point problems, singular perturbations, much more. Problems. 384pp. 5⅜ × 8½. 65456-7 Pa. $8.95

AMATEUR ASTRONOMER'S HANDBOOK, J.B. Sidgwick. Timeless, comprehensive coverage of telescopes, mirrors, lenses, mountings, telescope drives, micrometers, spectroscopes, more. 189 illustrations. 576pp. 5⅜ × 8¼. 24034-7 Pa. $8.95

CATALOG OF DOVER BOOKS

SPECIAL FUNCTIONS, N.N. Lebedev. Translated by Richard Silverman. Famous Russian work treating more important special functions, with applications to specific problems of physics and engineering. 38 figures. 308pp. 5⅜ × 8½.
60624-4 Pa. $6.95

OBSERVATIONAL ASTRONOMY FOR AMATEURS, J.B. Sidgwick. Mine of useful data for observation of sun, moon, planets, asteroids, aurorae, meteors, comets, variables, binaries, etc. 39 illustrations 384pp. 5⅜ × 8¼. (Available in U.S. only)
24033-9 Pa. $5.95

INTEGRAL EQUATIONS, F.G. Tricomi. Authoritative, well-written treatment of extremely useful mathematical tool with wide applications. Volterra Equations, Fredholm Equations, much more. Advanced undergraduate to graduate level. Exercises. Bibliography. 238pp. 5⅜ × 8½.
64828-1 Pa. $6.95

CELESTIAL OBJECTS FOR COMMON TELESCOPES, T.W. Webb. Inestimable aid for locating and identifying nearly 4,000 celestial objects. 77 illustrations. 645pp. 5⅜ × 8½.
20917-2, 20918-0 Pa., Two-vol. set $12.00

MODERN NONLINEAR EQUATIONS, Thomas L. Saaty. Emphasizes practical solution of problems; covers seven types of equations. ". . . a welcome contribution to the existing literature. . . ."—*Math Reviews*. 490pp. 5⅜ × 8½. 64232-1 Pa. $9.95

FUNDAMENTALS OF ASTRODYNAMICS, Roger Bate et al. Modern approach developed by U.S. Air Force Academy. Designed as a first course. Problems, exercises. Numerous illustrations. 455pp. 5⅜ × 8½. 60061-0 Pa. $8.95

INTRODUCTION TO LINEAR ALGEBRA AND DIFFERENTIAL EQUATIONS, John W. Dettman. Excellent text covers complex numbers, determinants, orthonormal bases, Laplace transforms, much more. Exercises with solutions. Undergraduate level. 416pp. 5⅜ × 8½. 65191-6 Pa. $8.95

INCOMPRESSIBLE AERODYNAMICS, edited by Bryan Thwaites. Covers theoretical and experimental treatment of the uniform flow of air and viscous fluids past two-dimensional aerofoils and three-dimensional wings; many other topics. 654pp. 5⅜ × 8½. 65465-6 Pa. $14.95

INTRODUCTION TO DIFFERENCE EQUATIONS, Samuel Goldberg. Exceptionally clear exposition of important discipline with applications to sociology, psychology, economics. Many illustrative examples; over 250 problems. 260pp. 5⅜ × 8½. 65084-7 Pa. $6.95

LAMINAR BOUNDARY LAYERS, edited by L. Rosenhead. Engineering classic covers steady boundary layers in two- and three-dimensional flow, unsteady boundary layers, stability, observational techniques, much more. 708pp. 5⅜ × 8½.
65646-2 Pa. $15.95

LECTURES ON CLASSICAL DIFFERENTIAL GEOMETRY, Second Edition, Dirk J. Struik. Excellent brief introduction covers curves, theory of surfaces, fundamental equations, geometry on a surface, conformal mapping, other topics. Problems. 240pp. 5⅜ × 8½. 65609-8 Pa. $6.95

GEOMETRY OF COMPLEX NUMBERS, Hans Schwerdtfeger. Illuminating, widely praised book on analytic geometry of circles, the Moebius transformation, and two-dimensional non-Euclidean geometries. 200pp. 5⅜ × 8¼.
63830-8 Pa. $6.95

MECHANICS, J.P. Den Hartog. A classic introductory text or refresher. Hundreds of applications and design problems illuminate fundamentals of trusses, loaded beams and cables, etc. 334 answered problems. 462pp. 5⅜ × 8½. 60754-2 Pa. $8.95

TOPOLOGY, John G. Hocking and Gail S. Young. Superb one-year course in classical topology. Topological spaces and functions, point-set topology, much more. Examples and problems. Bibliography. Index. 384pp. 5⅜ × 8¼.
65676-4 Pa. $7.95

STRENGTH OF MATERIALS, J.P. Den Hartog. Full, clear treatment of basic material (tension, torsion, bending, etc.) plus advanced material on engineering methods, applications. 350 answered problems. 323pp. 5⅜ × 8½. 60755-0 Pa. $7.50

ELEMENTARY CONCEPTS OF TOPOLOGY, Paul Alexandroff. Elegant, intuitive approach to topology from set-theoretic topology to Betti groups; how concepts of topology are useful in math and physics. 25 figures. 57pp. 5⅜ × 8½.
60747-X Pa. $2.95

ADVANCED STRENGTH OF MATERIALS, J.P. Den Hartog. Superbly written advanced text covers torsion, rotating disks, membrane stresses in shells, much more. Many problems and answers. 388pp. 5⅜ × 8½. 65407-9 Pa. $8.95

COMPUTABILITY AND UNSOLVABILITY, Martin Davis. Classic graduate-level introduction to theory of computability, usually referred to as theory of recurrent functions. New preface and appendix. 288pp. 5⅜ × 8½. 61471-9 Pa. $6.95

GENERAL CHEMISTRY, Linus Pauling. Revised 3rd edition of classic first-year text by Nobel laureate. Atomic and molecular structure, quantum mechanics, statistical mechanics, thermodynamics correlated with descriptive chemistry. Problems. 992pp. 5⅜ × 8½. 65622-5 Pa. $18.95

AN INTRODUCTION TO MATRICES, SETS AND GROUPS FOR SCIENCE STUDENTS, G. Stephenson. Concise, readable text introduces sets, groups, and most importantly, matrices to undergraduate students of physics, chemistry, and engineering. Problems. 164pp. 5⅜ × 8½. 65077-4 Pa. $5.95

THE HISTORICAL BACKGROUND OF CHEMISTRY, Henry M. Leicester. Evolution of ideas, not individual biography. Concentrates on formulation of a coherent set of chemical laws. 260pp. 5⅜ × 8½. 61053-5 Pa. $6.00

THE PHILOSOPHY OF MATHEMATICS: An Introductory Essay, Stephan Körner. Surveys the views of Plato, Aristotle, Leibniz & Kant concerning propositions and theories of applied and pure mathematics. Introduction. Two appendices. Index. 198pp. 5⅜ × 8½. 25048-2 Pa. $5.95

THE DEVELOPMENT OF MODERN CHEMISTRY, Aaron J. Ihde. Authoritative history of chemistry from ancient Greek theory to 20th-century innovation. Covers major chemists and their discoveries. 209 illustrations. 14 tables. Bibliographies. Indices. Appendices. 851pp. 5⅜ × 8½. 64235-6 Pa. $15.95

TENSOR CALCULUS, J.L. Synge and A. Schild. Widely used introductory text covers spaces and tensors, basic operations in Riemannian space, non-Riemannian spaces, etc. 324pp. 5⅜ × 8¼. 63612-7 Pa. $7.00

A CONCISE HISTORY OF MATHEMATICS, Dirk J. Struik. The best brief history of mathematics. Stresses origins and covers every major figure from ancient Near East to 19th century. 41 illustrations. 195pp. 5⅜ × 8½. 60255-9 Pa. $7.95

A SHORT ACCOUNT OF THE HISTORY OF MATHEMATICS, W.W. Rouse Ball. One of clearest, most authoritative surveys from the Egyptians and Phoenicians through 19th-century figures such as Grassman, Galois, Riemann. Fourth edition. 522pp. 5⅜ × 8½. 20630-0 Pa. $9.95

HISTORY OF MATHEMATICS, David E. Smith. Non-technical survey from ancient Greece and Orient to late 19th century; evolution of arithmetic, geometry, trigonometry, calculating devices, algebra, the calculus. 362 illustrations. 1,355pp. 5⅜ × 8½. 20429-4, 20430-8 Pa., Two-vol. set $21.90

THE GEOMETRY OF RENÉ DESCARTES, René Descartes. The great work founded analytical geometry. Original French text, Descartes' own diagrams, together with definitive Smith-Latham translation. 244pp. 5⅜ × 8½. 60068-8 Pa. $6.00

THE ORIGINS OF THE INFINITESIMAL CALCULUS, Margaret E. Baron. Only fully detailed and documented account of crucial discipline: origins; development by Galileo, Kepler, Cavalieri; contributions of Newton, Leibniz, more. 304pp. 5⅜ × 8½. (Available in U.S. and Canada only) 65371-4 Pa. $7.95

THE HISTORY OF THE CALCULUS AND ITS CONCEPTUAL DEVELOPMENT, Carl B. Boyer. Origins in antiquity, medieval contributions, work of Newton, Leibniz, rigorous formulation. Treatment is verbal. 346pp. 5⅜ × 8½. 60509-4 Pa. $6.95

THE THIRTEEN BOOKS OF EUCLID'S ELEMENTS, translated with introduction and commentary by Sir Thomas L. Heath. Definitive edition. Textual and linguistic notes, mathematical analysis. 2500 years of critical commentary. Not abridged. 1,414pp. 5⅜ × 8½. 60088-2, 60089-0, 60090-4 Pa., Three-vol. set $26.85

A HISTORY OF VECTOR ANALYSIS: The Evolution of the Idea of a Vectorial System, Michael J. Crowe. The first large-scale study of the history of vector analysis, now the standard on the subject. Unabridged republication of the edition published by University of Notre Dame Press, 1967, with second preface by Michael C. Crowe. Index. 278pp. 5⅜ × 8½. 64955-5 Pa. $7.00

THE HISTORICAL ROOTS OF ELEMENTARY MATHEMATICS, Lucas N.H. Bunt, Phillip S. Jones, and Jack D. Bedient. Fundamental underpinnings of modern arithmetic, algebra, geometry and number systems derived from ancient civilizations. 320pp. 5⅜ × 8½. 25563-8 Pa. $7.95

CALCULUS REFRESHER FOR TECHNICAL PEOPLE, A. Albert Klaf. Covers important aspects of integral and differential calculus via 756 questions. 566 problems, most answered. 431pp. 5⅜ × 8½. 20370-0 Pa. $7.95

THE FOUR-COLOR PROBLEM: Assaults and Conquest, Thomas L. Saaty and Paul G. Kainen. Engrossing, comprehensive account of the century-old combinatorial topological problem, its history and solution. Bibliographies. Index. 110 figures. 228pp. 5⅜ × 8½. 65092-8 Pa. $6.00

CATALYSIS IN CHEMISTRY AND ENZYMOLOGY, William P. Jencks. Exceptionally clear coverage of mechanisms for catalysis, forces in aqueous solution, carbonyl- and acyl-group reactions, practical kinetics, more. 864pp. 5⅜ × 8½. 65460-5 Pa. $18.95

PROBABILITY: An Introduction, Samuel Goldberg. Excellent basic text covers set theory, probability theory for finite sample spaces, binomial theorem, much more. 360 problems. Bibliographies. 322pp. 5⅜ × 8½. 65252-1 Pa. $7.95

LIGHTNING, Martin A. Uman. Revised, updated edition of classic work on the physics of lightning. Phenomena, terminology, measurement, photography, spectroscopy, thunder, more. Reviews recent research. Bibliography. Indices. 320pp. 5⅜ × 8¼. 64575-4 Pa. $7.95

PROBABILITY THEORY: A Concise Course, Y.A. Rozanov. Highly readable, self-contained introduction covers combination of events, dependent events, Bernoulli trials, etc. Translation by Richard Silverman. 148pp. 5⅜ × 8¼. 63544-9 Pa. $4.50

THE CEASELESS WIND: An Introduction to the Theory of Atmospheric Motion, John A. Dutton. Acclaimed text integrates disciplines of mathematics and physics for full understanding of dynamics of atmospheric motion. Over 400 problems. Index. 97 illustrations. 640pp. 6 × 9. 65096-0 Pa. $16.95

STATISTICS MANUAL, Edwin L. Crow, et al. Comprehensive, practical collection of classical and modern methods prepared by U.S. Naval Ordnance Test Station. Stress on use. Basics of statistics assumed. 288pp. 5⅜ × 8½. 60599-X Pa. $6.00

WIND WAVES: Their Generation and Propagation on the Ocean Surface, Blair Kinsman. Classic of oceanography offers detailed discussion of stochastic processes and power spectral analysis that revolutionized ocean wave theory. Rigorous, lucid. 676pp. 5⅜ × 8½. 64652-1 Pa. $14.95

STATISTICAL METHOD FROM THE VIEWPOINT OF QUALITY CONTROL, Walter A. Shewhart. Important text explains regulation of variables, uses of statistical control to achieve quality control in industry, agriculture, other areas. 192pp. 5⅜ × 8½. 65232-7 Pa. $6.00

THE INTERPRETATION OF GEOLOGICAL PHASE DIAGRAMS, Ernest G. Ehlers. Clear, concise text emphasizes diagrams of systems under fluid or containing pressure; also coverage of complex binary systems, hydrothermal melting, more. 288pp. 6½ × 9¼. 65389-7 Pa. $8.95

STATISTICAL ADJUSTMENT OF DATA, W. Edwards Deming. Introduction to basic concepts of statistics, curve fitting, least squares solution, conditions without parameter, conditions containing parameters. 26 exercises worked out. 271pp. 5⅜ × 8½. 64685-8 Pa. $7.95

DE RE METALLICA, Georgius Agricola. The famous Hoover translation of greatest treatise on technological chemistry, engineering, geology, mining of early modern times (1556). All 289 original woodcuts. 638pp. 6¾ × 11.
60006-8 Clothbd. $15.95

SOME THEORY OF SAMPLING, William Edwards Deming. Analysis of the problems, theory and design of sampling techniques for social scientists, industrial managers and others who find statistics increasingly important in their work. 61 tables. 90 figures. xvii + 602pp. 5⅜ × 8½.
64684-X Pa. $14.95

THE VARIOUS AND INGENIOUS MACHINES OF AGOSTINO RAMELLI: A Classic Sixteenth-Century Illustrated Treatise on Technology, Agostino Ramelli. One of the most widely known and copied works on machinery in the 16th century. 194 detailed plates of water pumps, grain mills, cranes, more. 608pp. 9 × 12.
25497-6 Clothbd. $34.95

LINEAR PROGRAMMING AND ECONOMIC ANALYSIS, Robert Dorfman, Paul A. Samuelson and Robert M. Solow. First comprehensive treatment of linear programming in standard economic analysis. Game theory, modern welfare economics, Leontief input-output, more. 525pp. 5⅜ × 8½.
65491-5 Pa. $12.95

ELEMENTARY DECISION THEORY, Herman Chernoff and Lincoln E. Moses. Clear introduction to statistics and statistical theory covers data processing, probability and random variables, testing hypotheses, much more. Exercises. 364pp. 5⅜ × 8½.
65218-1 Pa. $8.95

THE COMPLEAT STRATEGYST: Being a Primer on the Theory of Games of Strategy, J.D. Williams. Highly entertaining classic describes, with many illustrated examples, how to select best strategies in conflict situations. Prefaces. Appendices. 268pp. 5⅜ × 8½.
25101-2 Pa. $5.95

MATHEMATICAL METHODS OF OPERATIONS RESEARCH, Thomas L. Saaty. Classic graduate-level text covers historical background, classical methods of forming models, optimization, game theory, probability, queueing theory, much more. Exercises. Bibliography. 448pp. 5⅜ × 8¼.
65703-5 Pa. $12.95

CONSTRUCTIONS AND COMBINATORIAL PROBLEMS IN DESIGN OF EXPERIMENTS, Damaraju Raghavarao. In-depth reference work examines orthogonal Latin squares, incomplete block designs, tactical configuration, partial geometry, much more. Abundant explanations, examples. 416pp. 5⅜ × 8¼.
65685-3 Pa. $10.95

THE ABSOLUTE DIFFERENTIAL CALCULUS (CALCULUS OF TENSORS), Tullio Levi-Civita. Great 20th-century mathematician's classic work on material necessary for mathematical grasp of theory of relativity. 452pp. 5⅜ × 8½.
63401-9 Pa. $9.95

VECTOR AND TENSOR ANALYSIS WITH APPLICATIONS, A.I. Borisenko and I.E. Tarapov. Concise introduction. Worked-out problems, solutions, exercises. 257pp. 5⅜ × 8¼.
63833-2 Pa. $6.95

CATALOG OF DOVER BOOKS

CHALLENGING MATHEMATICAL PROBLEMS WITH ELEMENTARY SOLUTIONS, A.M. Yaglom and I.M. Yaglom. Over 170 challenging problems on probability theory, combinatorial analysis, points and lines, topology, convex polygons, many other topics. Solutions. Total of 445pp. 5⅜ × 8½. Two-vol. set.

Vol. I 65536-9 Pa. $5.95
Vol. II 65537-7 Pa. $5.95

FIFTY CHALLENGING PROBLEMS IN PROBABILITY WITH SOLUTIONS, Frederick Mosteller. Remarkable puzzlers, graded in difficulty, illustrate elementary and advanced aspects of probability. Detailed solutions. 88pp. 5⅜ × 8½.
65355-2 Pa. $3.95

EXPERIMENTS IN TOPOLOGY, Stephen Barr. Classic, lively explanation of one of the byways of mathematics. Klein bottles, Moebius strips, projective planes, map coloring, problem of the Koenigsberg bridges, much more, described with clarity and wit. 43 figures. 210pp. 5⅜ × 8½.
25933-1 Pa. $4.95

RELATIVITY IN ILLUSTRATIONS, Jacob T. Schwartz. Clear non-technical treatment makes relativity more accessible than ever before. Over 60 drawings illustrate concepts more clearly than text alone. Only high school geometry needed. Bibliography. 128pp. 6⅛ × 9¼.
25965-X Pa. $5.95

AN INTRODUCTION TO ORDINARY DIFFERENTIAL EQUATIONS, Earl A. Coddington. A thorough and systematic first course in elementary differential equations for undergraduates in mathematics and science, with many exercises and problems (with answers). Index. 304pp. 5⅜ × 8¼.
65942-9 Pa. $7.95

FOURIER SERIES AND ORTHOGONAL FUNCTIONS, Harry F. Davis. An incisive text combining theory and practical example to introduce Fourier series, orthogonal functions and applications of the Fourier method to boundary-value problems. 570 exercises. Answers and notes. 416pp. 5⅜ × 8½. 65973-9 Pa. $8.95

THE THOERY OF BRANCHING PROCESSES, Theodore E. Harris. First systematic, comprehensive treatment of branching (i.e. multiplicative) processes and their applications. Galton-Watson model, Markov branching processes, electron-photon cascade, many other topics. Rigorous proofs. Bibliography. 240pp. 5⅜ × 8½.
65952-6 Pa. $6.95

AN INTRODUCTION TO ALGEBRAIC STRUCTURES, Joseph Landin. Superb self-contained text covers "abstract algebra": sets and numbers, theory of groups, theory of rings, much more. Numerous well-chosen examples, exercises. 247pp. 5⅜ × 8½.
65940-2 Pa. $6.95

GAMES AND DECISIONS: Introduction and Critical Survey, R. Duncan Luce and Howard Raiffa. Superb non-technical introduction to game theory, primarily applied to social sciences. Utility theory, zero-sum games, n-person games, decision-making, much more. Bibliography. 509pp. 5⅜ × 8½. 65943-7 Pa. $10.95

CATALOG OF DOVER BOOKS

ROTARY-WING AERODYNAMICS, W.Z. Stepniewski. Clear, concise text covers aerodynamic phenomena of the rotor and offers guidelines for helicopter performance evaluation. Originally prepared for NASA. 537 figures. 640pp. 6⅛ × 9¼.
64647-5 Pa. $14.95

DIFFERENTIAL GEOMETRY, Heinrich W. Guggenheimer. Local differential geometry as an application of advanced calculus and linear algebra. Curvature, transformation groups, surfaces, more. Exercises. 62 figures. 378pp. 5⅜ × 8½.
63433-7 Pa. $7.95

INTRODUCTION TO SPACE DYNAMICS, William Tyrrell Thomson. Comprehensive, classic introduction to space-flight engineering for advanced undergraduate and graduate students. Includes vector algebra, kinematics, transformation of coordinates. Bibliography. Index. 352pp. 5⅜ × 8½.   65113-4 Pa. $8.00

A SURVEY OF MINIMAL SURFACES, Robert Osserman. Up-to-date, in-depth discussion of the field for advanced students. Corrected and enlarged edition covers new developments. Includes numerous problems. 192pp. 5⅜ × 8½.
64998-9 Pa. $8.00

ANALYTICAL MECHANICS OF GEARS, Earle Buckingham. Indispensable reference for modern gear manufacture covers conjugate gear-tooth action, gear-tooth profiles of various gears, many other topics. 263 figures. 102 tables. 546pp. 5⅜ × 8½.   65712-4 Pa. $11.95

SET THEORY AND LOGIC, Robert R. Stoll. Lucid introduction to unified theory of mathematical concepts. Set theory and logic seen as tools for conceptual understanding of real number system. 496pp. 5⅜ × 8¼.   63829-4 Pa. $8.95

A HISTORY OF MECHANICS, René Dugas. Monumental study of mechanical principles from antiquity to quantum mechanics. Contributions of ancient Greeks, Galileo, Leonardo, Kepler, Lagrange, many others. 671pp. 5⅜ × 8½.
65632-2 Pa. $14.95

FAMOUS PROBLEMS OF GEOMETRY AND HOW TO SOLVE THEM, Benjamin Bold. Squaring the circle, trisecting the angle, duplicating the cube: learn their history, why they are impossible to solve, then solve them yourself. 128pp. 5⅜ × 8½.   24297-8 Pa. $3.95

MECHANICAL VIBRATIONS, J.P. Den Hartog. Classic textbook offers lucid explanations and illustrative models, applying theories of vibrations to a variety of practical industrial engineering problems. Numerous figures. 233 problems, solutions. Appendix. Index. Preface. 436pp. 5⅜ × 8½.   64785-4 Pa. $8.95

CURVATURE AND HOMOLOGY, Samuel I. Goldberg. Thorough treatment of specialized branch of differential geometry. Covers Riemannian manifolds, topology of differentiable manifolds, compact Lie groups, other topics. Exercises. 315pp. 5⅜ × 8½.   64314-X Pa. $6.95

HISTORY OF STRENGTH OF MATERIALS, Stephen P. Timoshenko. Excellent historical survey of the strength of materials with many references to the theories of elasticity and structure. 245 figures. 452pp. 5⅜ × 8½.   61187-6 Pa. $9.95